ZAMM publishes original papers (main articles and short communications), the lectures presented at the annual GAMM conferences, and review articles (requested by the Editorial Office of ZAMM). Papers in English (preferably) or German to be submitted for publication should be sent to the address at the bottom of this page (master and one copy), in the case of original papers together with an explicit assurance that they have not been published previously in, or are submitted to, another public journal. Authors are held fully responsible for all information included in their paper and should have one copy available for correction purposes. They get two sets of page proofs, one of them has to be returned within one week. Authors may be charged for extensive corrections in the text; corrections in the figures are not possible. They receive 150 (plenary GAMM lectures), 75 (main articles and short GAMM lectures), or 25 (short communications), respectively, free-of-charge reprints. Additional reprints have to be paid for.

Manuscripts should be type-written (one-sided, double-spaced with wide margins at both sides, each sheet numbered) beginning with the authors' names and the title of the paper. Extra types (like bold-faced letters etc.) should be marked. Please, distinguish between 1, *l*, e; v, ν; w, ω; o, *O*, 0 (zero); *k*, *K*, $\varkappa$; *i*, ι; *x*, × (cross), χ etc. **Formulas** must be written clearly legible (type-writer if possible, never use pencil). Formulas used in the text have to be numbered at the right margin. Use abbreviations for long terms, "exp" instead of "e" for extensive exponents, bold-faced letters for vectors etc. instead of arrows; avoid hierarchies of indices and exponents if possible. The end of a proof should be marked.

Figures should be submitted on separate sheets, drawn with Indian ink and numbered consecutively. Figure captions, tables, and footnotes (if not avoidable) should be inserted in the final place in the text.

References should be numbered in brackets throughout the text and collected according to these numbers in the appended bibliography (References). Each reference should contain author(s), title, and complete bibliographical coordinates following the subsequent examples.

Books (Reports, Preprints etc.):
Irons, B. M.; Shrive N. G.: Numerical methods in engineering and applied science — Numbers are fun. Ellis Horwood Series in Mathematics and Its Applications, Ellis Horwood Ltd., Chichester 1987.

Articles in periodicals:
Kassem, S. E.: Wave source potentials for two superposed fluids, each of finite depth. Indian J. Pure Appl. Math. **18** (1987) 2, 186—193.

Articles in Proceedings etc.:
Labisch, F. K.: Some remarks on the morphology of non-unique solutions in nonlinear elastostatics. In: Küpper, T.; Seydel, R.; Troger, H. (eds.): Bifurcation: Analysis, algorithms, applications. Proc. Conf. at the Univ. of Dortmund, August 18—22, 1986. Internat. Ser. of Numer. Mathematics, Vol. 79, Birkhäuser Verlag, Basel-Boston-Stuttgart 1987, pp. 177—184.

To each original paper and review article, a **summary** of 5 to 10 lines in English (for manuscripts of more than 10 pages in German and Russian) and a short list of key-words and corresponding AMS (MOS) classification numbers are needed.

The precise address of each author with full name, academic title etc. is to be added. Information on address changes is appreciated.

ZAMM veröffentlicht Originalarbeiten (Hauptaufsätze und Kleine Mitteilungen), die Vorträge der jährlichen GAMM-Tagungen und Übersichtsartikel (die von der Redaktion erbeten werden). Arbeiten in Englisch (vorzugsweise) oder Deutsch, die zur Publikation eingereicht werden, sind an die unten stehende Adresse zu senden (Original und eine Kopie). Dem Manuskript von Originalarbeiten ist eine Erklärung beizufügen, daß die Arbeit bisher in keiner anderen Zeitschrift veröffentlicht worden ist und auch kein Antrag auf Veröffentlichung in einer solchen Zeitschrift läuft. Die Autoren sind für den Inhalt ihrer Arbeiten voll verantwortlich und sollten für Korrekturzwecke eine weitere Kopie besitzen. Sie erhalten die Korrekturfahnen in zweifacher Ausfertigung; eine Korrektur ist innerhalb einer Woche zurückzusenden. Die Kosten für umfangreiche nachträgliche Korrekturen können dem Autor berechnet werden. Korrekturen in den Abbildungen sind nicht möglich. Die Autoren erhalten kostenlos 150 Sonderdrucke für die Hauptvorträge der GAMM-Tagungen, 75 für Hauptaufsätze, Übersichtsartikel und die GAMM-Kurzvorträge und 25 für Kleine Mitteilungen. Zusätzliche Sonderdrucke sind gegen Berechnung erhältlich.

Die **Manuskripte** müssen mit Maschine geschrieben sein (einseitig, zweizeiliger Zeilenabstand mit breitem Rand auf beiden Seiten, jede Seite numeriert), wobei mit den Namen der Autoren begonnen werden soll und danach der Titel folgt. Besondere Typen (z. B. halbfett) sind zu markieren. Es soll gut unterschieden werden zwischen 1, *l*, e; v, ν; w, ω, o, *O*, 0 (Null); *k*, *K*, $\varkappa$; *i*, ι; *x*, × (Kreuz), χ; usw. Die **Formeln** müssen deutlich lesbar sein (möglichst maschinenschriftlich, keine Verwendung von Bleistift). Im Text verwendete Formeln sollten am rechten Rand durchnumeriert werden. Für lange Terme sind Abkürzungen zu verwenden, „exp" anstelle von „e" für große Ausdrücke im Exponenten, halbfett für Vektoren anstelle von darübergesetzten Pfeilen — Mehrfach-Indizes wie auch -Exponenten möglichst vermeiden. Das Ende eines Beweises sollte gekennzeichnet werden.

Abbildungen sind auf gesonderten Blättern einzureichen und sollten mit Tusche gezeichnet und durchnumeriert sein. Die Abbildungsunterschriften, eventuelle Tabellen und (falls nicht vermeidbar) Fußnoten sind an der Stelle einzufügen, an der sie im Text erscheinen sollen.

Literaturzitate sind durch den gesamten Text durchzunumerieren (in eckigen Klammern) und in der entsprechenden Reihenfolge im am Schluß folgenden Literaturverzeichnis aufzulisten. Jedes Zitat sollte Autor(en), Titel und die vollständigen bibliografischen Koordinaten enthalten, entsprechend den folgenden Beispielen.

Bücher (Reports, Preprints usw.):
Irons, B. M.; Shrive N. G.: Numerical methods in engineering and applied science — Numbers are fun. Ellis Horwood Series in Mathematics and Its Applications, Ellis Horwood Ltd., Chichester 1987.

Zeitschriftenartikel:
Kassem, S. E.: Wave source potentials for two superposed fluids, each of finite depth. Indian J. Pure Appl. Math. **18** (1987) 2, 186—193.

Arbeiten in Proceedings usw.:
Labisch, F. K.: Some remarks on the morphology of non-unique solutions in nonlinear elastostatics. In: Küpper, T.; Seydel, R.; Troger, H. (eds.): Bifurcation: Analysis, algorithms, applications. Proc. Conf. at the Univ. of Dortmund, August 18—22, 1986. Internat. Ser. of Numer. Mathematics, Vol. 79, Birkhäuser Verlag, Basel-Boston-Stuttgart 1987, pp. 777—184.

Für jede Originalarbeit und jeden Übersichtsartikel wird eine englische **Zusammenfassung** (bei Manuskripten von mehr als 10 Seiten auch je eine in Deutsch und Russisch) von 5 bis 10 Zeilen Länge sowie eine kurze Liste von Stichworten und den entsprechenden AMS (MOS)-Klassifikationsmarken benötigt.

Die genaue Adresse jedes Autors mit vollständigem Namen, akademischen Titeln usw. muß beigefügt werden. Bei Adressenänderung wird eine umgehende Information erwartet.

ZAMM
Editorial Office
Rudower Chaussee 5
O-1199 Berlin

Zeitschrift für Angewandte Mathematik und Mechanik

ZAMM

Applied Mathematics and Mechanics

Founded by Richard von Mises in 1921

Edited at Institute of Mechanics
Academy of Sciences of the G.D.R.

Akademie-Verlag · Berlin

Volume 70 · 1990
Number 11

Editorial Office

Academy of Sciences of the GDR, Institute of Mechanics
Prof. Dr. Günter Schmidt · Elke Herrmann · Dr. Dieter Ullmann · Dr. Horst Weinert

The aim and scope of ZAMM is, in agreement with the intentions of its founder Richard von Mises, one of the outstanding scientists of our century in both Mathematics and Mechanics: to publish new results and review articles, the proceedings of the annual GAMM conferences, book reviews and information on applied mathematics (mainly on numerical mathematics and various parts and applications of analysis, in particular numerical aspects of differential and integral equations), on the whole field of theoretical and applied mechanics (solid mechanics, fluid mechanics, thermodynamics) and on mathematical physics.
The journal is of interest for persons working in applied mathematics and mechanics as well as on mathematical or mechanical questions in other sciences, for instance mechanical and civil engineering, electrotechniques, and chemistry.

In Übereinstimmung mit den Zielstellungen des Begründers der ZAMM, Richard von Mises, eines der bedeutendsten Wissenschaftler unseres Jahrhunderts auf dem Gebiet der Mathematik und Mechanik, publiziert die Zeitschrift neue Ergebnisse und Übersichtsartikel, die Vorträge auf den jährlichen GAMM-Tagungen, Buchbesprechungen und Informationen über Angewandte Mathematik (hauptsächlich über numerische Mathematik und verschiedene Teile sowie Anwendungen der Analysis, insbesondere numerische Aspekte bei Differential- und Integralgleichungen), über das gesamte Gebiet der theoretischen und angewandten Mechanik (Festkörpermechanik, Strömungsmechanik, Thermodynamik) und über mathematische Physik.
Die Zeitschrift wendet sich an Vertreter der angewandten Mathematik und Mechanik wie auch an Wissenschaftler und Ingenieure, die sich mit mathematischen oder theoretisch-mechanischen Fragen auf anderen Gebieten, z. B. Maschinenbau und Technik, Elektrotechnik und Chemie, befassen.

Terms of subscription

Orders can be sent

- **in the GDR:** to Postzeitungsvertrieb, or to Akademie-Verlag Berlin, Leipziger Str. 3–4, PF-Nr. 1233, O-1086 Berlin;
- **in the FRG and Berlin (West):** to a book-shop or to the wholesale distributing agency Kunst und Wissen, Erich Bieber GmbH, a Faxon Company, Postfach 102844, W-7000 Stuttgart 10;
- **in other countries** to the international book and periodical trading or to the wholesale distributing agency KUNST UND WISSEN, Erich Bieber GmbH, a Faxon Company, Stampfenbachstr. 73, Postfach, CH-8035 Zürich, or to a bookshop for foreign language literature, or to the competent newspaper sales agency of the country resp. to the AKADEMIE-VERLAG BERLIN, Leipziger Str. 3–4, O-1086 Berlin.

Bezugsmöglichkeiten

Bestellungen sind zu richten:

- **in der DDR** an den Postzeitungsvertrieb oder an den Akademie-Verlag Berlin, Leipziger Str. 3–4, PF-Nr. 1233, O-1086 Berlin;
- **in der BRD und Berlin (West)** an eine Buchhandlung oder an die Auslieferungsstelle Kunst und Wissen, Erich Bieber GmbH, ein Unternehmen der Faxon-Gruppe, Postfach 102844 W-7000 Stuttgart 10;
- **in den übrigen Ländern** an den internationalen Buch- und Zeitschriftenhandel bzw. an die Auslieferungsstelle KUNST UND WISSEN, Erich Bieber GmbH, ein Unternehmen der Faxon-Gruppe, Stampfenbachstr. 73, Postfach, CH-8035 Zürich, oder an eine Buchhandlung für fremdsprachige Literatur, an den zuständigen Postzeitungsvertrieb des Landes bzw. an den AKADEMIE-VERLAG BERLIN, Leipziger Str. 3–4, O-1086 Berlin.

ZEITSCHRIFT FÜR ANGEWANDTE MATHEMATIK UND MECHANIK

Herausgeber: Institut für Mechanik, Chemnitz. Mit der Herausgabe beauftragt und Chefredakteur: Prof. Dr. Günter Schmidt, Rudower Chaussee 5, O-1199 Berlin, Fernruf: 6745611 oder 6743668 (gleichzeitig Anschrift der Redaktion).
Verlag: Akademie-Verlag Berlin, Leipziger Straße 3–4, PF-Nr. 1233, O-1086 Berlin; Fernruf: 2236221 oder 2236229; Telex-Nr.: 114420; Bank: Berliner Stadtbank AG, Kto.-Nr.: 6651-25-130512.
Verlagsdirektor: Dr. Bernhard Tesche; Stellv. Cheflektor Zeitschriften: Armin Beck; Redakteur der Abt. Zeitschriften: Marianne Simon.
Veröffentlicht unter der Registriernummer 1282.
Gesamtherstellung: Druckhaus „Thomas Müntzer" GmbH, O-5820 Bad Langensalza.
Erscheinungsweise: Die Zeitschrift für Angewandte Mathematik und Mechanik erscheint monatlich. Die 12 Hefte eines Jahres einschließlich Tagungshefte bilden einen Band. Bezugspreis je Band 480,– DM zuzüglich Versandspesen. Bezugspreis je Heft 40,– DM.
Bestellnummer dieses Heftes: 1009/70/11.

AN (EDV) 35937

ZAMM · Z. angew. Math. Mech. 70 (1990) 11, 479–488 Akademie-Verlag Berlin

THEJLL, P.; GILBERT, R. P.

The Use of MACSYMA for Solving Elliptic Boundary Value Problems

In der vorliegenden Arbeit stellen wir eine sogenannte Rand-Methode zur Lösung eines elliptischen Randwertproblems vor. Wie schon im Buch von HERRERA *[14] bemerkt wurde, gibt es zwei hauptsächliche Ansätze für die Formulierung von Randmethoden. Der eine basiert auf der Integralgleichungsmethode, der andere auf der Verwendung vollständiger Lösungssysteme. Wir werden uns dem zweiten zuwenden, da die Integralgleichungsmethode von* COLTON *und* GILBERT *[5] bzw.* GILBERT *und* LINZ *[12] behandelt wurde.*

In this work we shall present a so-called boundary method for solving an elliptic boundary value problem. As has been mentioned in the book of HERRERA *[14] there are two main approaches for the formulation of boundary methods. One is based on the method of integral equations and the other on the use of complete systems of solutions. We shall emphasize the second approach as the method of integral equations has been treated in* COLTON *and* GILBERT *[5] and* GILBERT *and* LINZ *[12].*

В настоящей работе предлагается так-называемый краевый метод для решения эллиптических краевых задач. Как упоминал Херрера в книге [14], существуют две главных подходов формулирования краевых методов. Один из них основывается на методе интегральных уравнений а другой на применении полных систем решений. Акцентироваем второй подход, потому что метод интегральных уравнений изучаться в работах Колтона и Гильберта [5] и Гильберта и Линца [12].

Introduction

For the purposes of this paper we find the notations of the above mentioned work of HERRERA to be very useful. Basically we shell present a method by which we may construct a complete system of solutions to a particular differential equation. For purposes of exposition we choose this equation to be

$$Qu := -\Delta u + P(r^2)\,u = 0\,, \qquad r := |x|\,,\; x \in \Omega \subset \mathbb{R}^n\,. \tag{1}$$

Moreover, we shall limit our discussion to the Dirichlet problem, even though our method might be adapted to treat other boundary value problems.

In order to construct a complete system of solutions for $Qu = 0$ in the domain $\Omega \subset \mathbb{R}^n$ we make use of the method of integral operators (cf. BERGMAN [1], GILBERT [9, 10, 11]). Furthermore, we shall assume that the Dirichlet data is sufficiently regular so that the traces $\gamma_0 u \in H^0(\Omega)$. ($H^s(\Omega)$, $H^s(\partial\Omega)$, as usual, will denote the Sobolev spaces of order s on Ω and $\partial\Omega$, respectively.) Indeed, we shall ask that the data is usually smooth, that Ω is simply connected, and the boundary $\partial\Omega$ is Liapunov.

Using the method of integral operators we are able to represent solutions of $Qu = 0$, i.e. functions $u \in N_Q$ (the null-space of Q) with functions $v \in N_P$, where $P := -\Delta$ by means of transmutation.

CARROLL [3, 4] defines a *transmutation* of P into Q as an operator B having the property that $BQ = PB$ where these are applied to suitable function spaces. For our particular choice of operators Q and P we may represent B in the form

$$(Bv)(x) := v(x) + \int_0^1 \sigma^{n-1} G(r, 1 - \sigma^2)\, v(x\sigma^2)\, d\sigma\,. \tag{2}$$

Moreover, the kernel $G(r, t)$ in (2) may be seen to satisfy the hyperbolic equation

$$2(1 - t)\, G_{rt} - G_r + r(G_{rr} - P(r^2)\, G) = 0\,, \tag{3}$$

along with the Goursat data

$$G(0, t) = 0, \qquad G(r, 0) = \int_0^r P(r^2)\, r\, dr\,. \tag{4}$$

It is possible to construct solutions to (3) and (4) using finite difference approximations for the Goursat problem (cf. GILBERT and LINZ [12]; however in the instance where $P(r^2)$ is analytic the G-function has a Taylor series expansion in the t-variable, namely

$$G(r, t) = \sum_{k=1}^{\infty} c_k(r^2)\, t^{k-1}\,. \tag{5}$$

The Taylor coefficients $c_k(r^2)$, moreover, may be computed recursively by the scheme

```
v(r) := integrate(r*p(r), r)                                   (6)
c[1]:v(r)
for i:1 thru m do(
c[i + 1]:1/(2*i)*integrate((-r*diff(c[i], r,2)
  + (2*i-1)*diff(c[i], r, 1)
+ r*p(r)*c[i]), r)
          )
define(g(r, t), sum(c[i]*t^(i-1), i, 1, m))
```

where integrate (a, b) means 'do the indefinite integral of a with respect to b', where define (a, b) means 'define a as the function b' and where diff (a, b, c) means 'take c derivatives of a with respect to b'.

It is an interesting fact that, for the case of $P(x) \in C(\bar{\Omega})$, the transmutation has an inverse B^{-1}, i.e.

$$B^{-1}PB = Q, \tag{7}$$

and that B^{-1} has a representation

$$B^{-1}u(x) = u(x) + \int_0^1 \sigma^{n-2} H(r, 1 - \sigma^2)\, u(x\sigma^2)\, d\sigma \tag{8}$$

where $H(r, t)$ satisfies a Goursat system similar to (3–4).

We refer to C^2-solutions of $Qu = 0$ as being in N_Q (the nullspace of the operator Q) and those of $Pv = 0$ as belonging to N_P. An interesting conclusion concerning the transmutation B is that $BN_P \to N_Q$ and $B^{-1}N_Q \to N_P$, i.e. B is a bijection. This is particularly useful for constructing solutions of boundary value problems, as we know that each element of N_Q and, in particular, that element u which takes on the prescribed boundary data, say

$$\textit{either} \quad \gamma_0 u = f_0(x) \quad \textit{or} \quad \gamma_1 u = f_1(x) \tag{9}$$

corresponds to a unique element $v \in N_P$. (Here γ_0 and γ_1 are the trace operators for the function u and its normal derivative respectively.)

Moreover, as the transmutation is a bijection on $C(\bar{\Omega})$ in the case of Dirichlet data ($C^1(\bar{\Omega})$ in the case of Neumann data) there exists also a $\psi \in C(\partial\Omega) \supset \gamma_0 v = \psi$. For the Neumann problem we would need to have more regularity imposed on $P(r^2)$, say $C^1(\bar{\Omega})$. However, for what follows we shall void all these concerns by simply assuming $P(r^2)$ is analytic. Indeed, the MACSYMA algorithm used requires this. (The computer language MACSYMA, cf. Bogen et al. [2], manipulates equations and other symbolic expressions. It was developed at the MIT Laboratory for Computer Science and was supported from 1975 to 1984 by NASA under grant NSG 1323, ONR under grant N0014-77-C-0641, by DOE under grant ET-78-C-02-4687 and by ASFOR under grant F49620-79-C-020. Since 1982, MACSYMA has been supported by Symbolics, Inc., who owns the trademark MACSYMA.)

Returning to the above argument it is clear that each solution of Dirichlet's problem for $Qu = 0$, $\gamma_0 u = f_0(x)$ corresponds to a unique solution of Dirichlet's problem for $Pv = 0$, $\gamma_0 v = \psi_0(x)$. We do not know off-hand what the data $\psi_0(x)$ is; however, what we do know is that this unique function transmutes into the desired solution of $Qu = 0$, $\gamma_0 u = f_0$. As it is known (cf. Vekua [17], Herrera [14]) that the solutions of $Pv = 0$, $\gamma_0 v = \psi_0$ may be approximated by harmonic polynomials when Ω is simply connected it is suggested that we seek our solutions to be prescribed problems as transmutations of these linear combinations of harmonic polynomials

$$v_N(x) := \sum_{k=1}^{N} \alpha_k B h_k(x), \tag{10}$$

where the $h_k(x)$ are harmonic polynomials and the α_k are constants.

Error estimates

As is well known (cf. Garabedian [7], Herrera [14]) the harmonic polynomials are a linearly independent, complete set of functions with respect to the norm induced by the Dirichlet inner product

$$(u, v) := \int_\Omega (\nabla u \cdot \nabla v)\, dx. \tag{11}$$

Suppose $H := \{h_1, h_2, \ldots\}$ represents the set of harmonic polynomials in $\mathbb{R}^n$. Then we construct the set $B := \{\varphi_1, \varphi_2, \ldots\}$, $\varphi_k(x) := (Bh_k)(x)$, of solutions to $Qu = 0$. We may show that B spans the spaces (cf. Herrera [14])

$$N_Q^{1/2} := \{u: u \in H^{1/2}(\Omega), \text{and } Qu = 0 \text{ in } \Omega\} \quad \text{and} \quad N_Q^{3/2} := \{u: u \in H^{3/2}(\Omega), \text{and } Qu = 0 \text{ in } \Omega\}.$$

For Dirichlet's problem we use the space $N_Q^{1/2}$. In order to approximate solutions to $Qu = 0$, $\gamma_0 u = \varphi_0$, we project the trace $\gamma_0 u$ onto the space B, that is we choose coefficients α_k^N such that the norm

$$\left\| \gamma_0 u - \sum_{k=1}^{N} \alpha_k^N \varphi_k \right\|^2_{L^2(\partial\Omega)} \tag{12}$$

is minimized. If we define the matrix coefficients

$$G_{ij} := (\varphi_i, \varphi_j)$$

where $(\varphi, \varphi) = \|\varphi\|^2_{L^2(\partial\Omega)}$, then the coefficients α_k^N are a solution of the linear system

$$\sum_{k=1}^{N} G_{kj}\alpha_k^N = d_j := \int_{\partial\Omega} \gamma_0 u \varphi_j d\sigma \tag{13}$$

where $d\sigma$ is the surface measure on $\partial\Omega$.

This approach leads to uniform approximation on compact subsets $K \in \Omega$; however, little may be said about convergence on $\bar{\Omega}$.

Transmutation, however, permits one to transform problems concerning the approximation on the boundary $\partial\Omega$ of solutions of $Qu = 0$ to the approximation on the boundary of the solutions of $Pv = 0$ (the harmonic functions) or what is essentially the same, using the Schwarz function, the approximation of an analytic function $\varphi(z)$. Concerning the interpolation problem for a bounded region in the plane we know a great deal. For example if $C := \partial\Omega$ and $\{z_0, \ldots, z_n\}$ are points on C where the analytic function $\varphi(z)$ takes on the corresponding values $\{w_0, \ldots, w_n\}$ then we know that a polynomial approximation exists that satisfies these values. It is given by GAIER [6],

$$L_n(z) := \sum_{k=0}^{n} w_k l_k(z), \tag{14}$$

with

$$l_k(z) := \frac{\omega(z)}{\omega'(z_k)(z - z_k)}, \quad k = 0, 1, \ldots, n, \quad \text{and} \quad \omega(z) = \prod_{k=0}^{n} (z - z_k).$$

If it is known that $\varphi(z)$ is analytic in a neighborhood of $\bar{\Omega}$ then we can get quite nice estimates concerning how the $L_n(z)$ converge uniformly on $\bar{\Omega}$ to $\varphi(z)$. Indeed, let Ω^C be the complement of Ω in $\mathbb{C}$ and $w = f(z)$ map Ω^C conformally on the exterior of the unit disk $\{w : |w| = 1\}$. If C_ϱ is the level line corresponding to the circle $|w| = \varrho > 1$ under the inverse conformal map $z = f^{-1}(w)$ which corresponds to the largest region in which $\varphi(z)$ is regular, then for interpolation polynomials L_n with uniformly distributed nodes in $\bar{\Omega}$ (or for that matter $\partial\Omega$) one has

$$\lim_{n\to\infty} (\max |\varphi(z) - L_n(z)|)^{1/n} = \frac{1}{\varrho}. \tag{15}$$

Unfortunately, $\varphi(z)$ may not be holomorphic in a neighborhood of $\bar{\Omega}$. However, the Mergelyan Theorem (1951, cf. GAIER [6]) gives us a partial answer to our problem. It states that if $\bar{\Omega}$ is compact and Ω^C connected, and if $\varphi(z)$ is holomorphic in Ω and continuous on $\bar{\Omega}$ then for each $\varepsilon > 0$ there exists a polynomial $P_n(z)$ such that $|\varphi(z) - P_n(z)| < \varepsilon/2$ for each $z \in \bar{\Omega}$. As the polynomial $P_n(z)$ is holomorphic in $\mathbb{C}$ it may be approximated by polynomials $L_{k_n}(z)$ uniformly on $\bar{\Omega}$ within an $\varepsilon/2$ by choosing k_n sufficiently large. Consequently, one obtains

$$|\varphi(z) - L_{k_n}(z)| \leqq |\varphi(z) - P_n(z)| + |P_n(z) - L_{k_n}(z)| < \varepsilon \quad \text{for } z \in \bar{\Omega}. \tag{16}$$

Hence, we may reduce the interpolation problem on $\partial\Omega$ for solutions of $QU = 0$ to finding the appropriate Lagrange polynomials for the associated analytic function $\varphi(z)$.

For example, let $u(x)$ be the actual, unique solution to the Dirichlet problem we have posed. In general, this will correspond to an infinite series made up from the complete family of solutions. To generate this complete family we have needed to compute the G-function which in general also consists of an infinite number of terms. In our case we have worked with an approximate G-function, G_N, consisting of a truncated series of N terms, and used this to construct n members of an approximate family. A function of this class we denote therefore by superscripts $u^{N,n}(x)$. A particular member of this class was chosen by the collocation method. From the triangle inequality we have that on the boundary C the difference between the solution $v(x)$ we would have constructed by taking the limits as $N \to \infty$ and $n \to \infty$ and the boundary data f is bounded by

$$\|v - f\|_C \leqq \|v(x) - u^{\infty,n}(x)\|_C + \|u^{\infty,n}(x) - u^{N,n}(x)\|_C + \|u^{N,n}(x) - f(x)\|_C \tag{17}$$

where $\|\ldots\|_C$ means the maximum norm on C. Clearly, the second term may be made small by taking N sufficiently large; moreover, the third term is made small by construction using the collocation approach. This may be shown to be always possible by an argumant using transmutation and Mergelyan's Theorem. Using the inverse transmutation we can estimate the first term by

$$\|v(x) - u^{\infty,n}(x)\|_C < \gamma \, \|h(x) - h^n(x)\|_C, \tag{18}$$

where $h(x)$ is the harmonic function that transmutes into $v(x)$ and $h^n(x)$ the harmonic polynomial that transmutes into $u^{\infty,n}(x)$. If $v(x)$ is the limit solution obtained by the processes of $N, n \to \infty$ it is also the collocation solution corresponding to fitting the complete family of solutions to an infinite number of collocation points. The transmuted harmonic function also corresponds to an associated collocation problem. From Mergelyan's theorem we may show that the collocation using the harmonic polynomials forces $h^n(x) \to h(x)$ on C. Hence $\|v - f\|_C < 3\varepsilon$, and as $\varepsilon > 0$ is arbitrary $v(x)$ is the actual solution to Dirichlet's problem. To compare $v(x)$ with the function actually computed we estimate

$$\|v(x) - u^{N,n}(x)\|_\Omega < \|v(x) - u^{\infty,n}(x)\|_\Omega + \|u^{N,n}(x) - u^{\infty,n}(x)\|_\Omega. \tag{19}$$

By the maximum principle the first term may be estimated, using transmutations, as

$$\|v(x) - u^{\infty,n}\|_\Omega \leqq \gamma \, \|h(x) - h^n(x)\|_\Omega < \varepsilon_1; \tag{20}$$

whereas, the second term may be estimated using

$$\begin{aligned} |u^{\infty,n}(x) - u^{N,n}(x)| &= \left| \sum_{k=N+1}^{\infty} \int_0^1 \sigma c_k(r^2)\, r(1 - \sigma^2)^{k-1} h^n(x\sigma^2) \, \mathrm{d}\sigma \right| \\ &\leqq \gamma \sum_{k=N+1}^{\infty} \int_0^1 \sigma |c_k(r^2)| \, (1 - \sigma^2)^N \, \mathrm{d}\sigma < \varepsilon_2, \qquad x \in \Omega, \end{aligned} \tag{21}$$

$\|h^n(x)\| \, 3 \leqq \gamma < \infty$ as for each n and the expansion on (5) converges in some disk $\Delta_R \Supset \Omega$. Consequently,

$$|v(x) - u^{N,n}(x)| < \varepsilon_1 + \varepsilon_2 \quad \text{for} \quad x \in \bar{\Omega}, \tag{22}$$

from which we see that the approximate solution is to have an error depending on the sum of the magnitude of the error in fitting the collocation and computing the G-function, as was expected.

Applications

The principle of transmuting the solutions of the homogeneous equation $\Delta u = 0$ into the solutions of (1), as discussed above was put into an interactive program that uses MACSYMA and FORTRAN 77 to find the G-function for (1) and then the coefficients α_i in (10). In 2 dimensions the polynomials ψ are:

$$\psi_j(r, \theta) = r^k \cos(k\theta), \; r^k \sin(k\theta) \tag{23}$$

for odd and even j, respectively; k is the integer part of $j/2$. Then

$$\varphi_j(r, \theta) = r^k \{\cos(k\theta) \text{ or } \sin(k\theta)\} \, (1 + m_k(r)) \tag{24}$$

where

$$m_k(r) = \int_0^1 \sigma^{2k+1} G(r, 1 - \sigma^2) \, \mathrm{d}\sigma. \tag{25}$$

The above forms the basis for the algorithm programmed in MACSYMA, which we present in this paper.

The α_i are found by fitting the approximate solution to a number of points along the boundary. The G-function and the φ's are found entirely within MACSYMA while the rest of the calculations in the scheme are done numerically with FORTRAN routines that are managed from MACSYMA.

The program

The MACSYMA program we have written is shown in Appendix A. It is made up of several functions, each with a specific task, which are called consecutively from the main function 'transmute'. The program is started by typing transmute (); and the program then asks the user for input. The function $P(r^2)$ and the number of terms wanted in the calculation of the G-function and in the sum over the φ_i's must be specified. When a G-function has been calculated it is saved so that it can be reused later. When the $P(r^2)$ function is not a simple constant the calculation time for the G-function can be large, especially if many terms are required. Then it becomes very useful to be able to reuse the same, stored, G-function, for later runs. The same can be done for the calculated set of φ_i's.

The form of the function on the boundary must also be given and so must the equation for the boundary. Since this program is for 2-dimensional problems it is natural to use plane polar coordinates. The radial coordinate must be specified as a function of the angular coordinate.

The program then proceeds to calculate the φ_i's. We choose to determine the coefficients α_i by fitting the approximate solution to a number of points along the boundary. To enable comparison with GILBERT and LINZ [12] we choose to fit at 31 points. For fewer than 31 basefunctions φ_i this will lead to an overdetermined system of equations which we solve in the least square sense using the IMSL routine LLSQF.

To do this we let our MACSYMA program write the FORTRAN 77 program needed, compile and run it and import the solutions (the α_i's) to the MACSYMA program. This feature of MACSYMA is very attractive and useful.

With the now known approximate solution we go on to estimate the fitting error in the same way as in GILBERT and LINZ [12]. The error is

$$\max \{|boundary\text{-}function(point_i) - approx\_solution(point_i)|\} \quad \text{for} \quad i = 1, \ldots, 31\,. \tag{26}$$

This step is also performed in FORTRAN 77 since the repeated numerical evaluation needed is a very time consuming process in MACSYMA.

The approximate solution found in each run is saved in a file called 'appsol' and can be reloaded at a later time.

Test runs

We tested our program on a selection of problems. Two problems were those of GILBERT and LINZ [12].

The problem $\Delta u - u = 0$, $u = \exp(\sin(\theta))$ on a circle of radius 1

These results of our calculations along with the results of GILBERT and LINZ [12] are shown in Fig. 1 below. The numerical calculations were performed in single precision FORTRAN 77 and this is probably the reason for the flattening out of the error as the number of terms in the sum (10) goes beyond 17. The error does not go down evenly as the number of

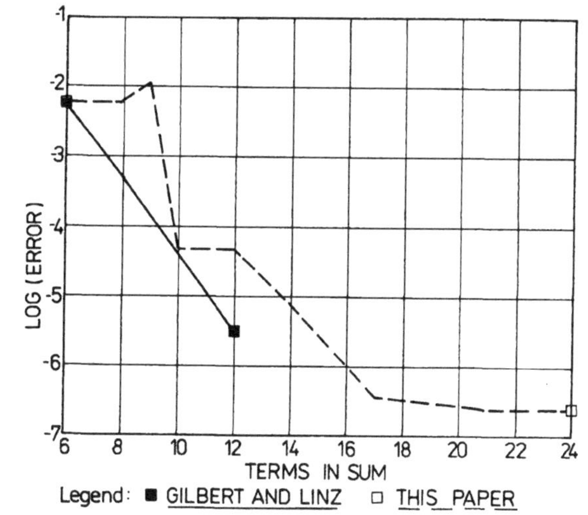

Fig. 1. The problem $\Delta u - u = 0$, $u = e^{\sin(\theta)}$ on a circle of radius 1. The numerical calculations were performed in single precision FORTRAN 77 and this is probably the reason for the flattening out of the error as the number of terms in the sum goes beyond 17. The error does not go down evenly as the number of terms increases and the reason for this is unknown. That the errors we get are not the same as those of GILBERT and LINZ [12] for $n = 8$ and 12 can be due to the way we choose the 31 points at which to calculate the error. We distributed the points evenly around the perimeter of the circle. The G-function was calculated with 11 terms.

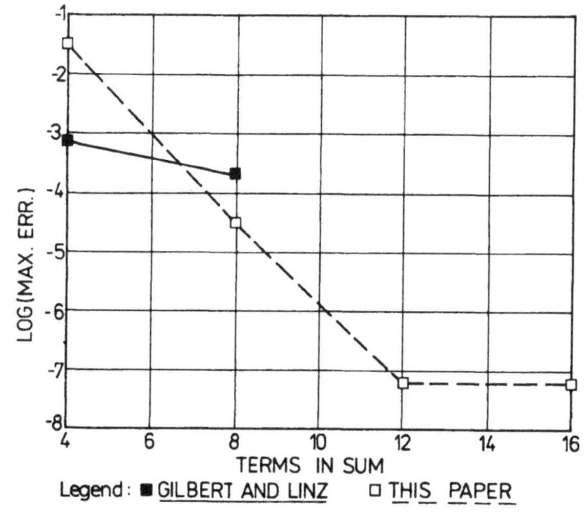

Fig. 2. The problem $\Delta u - u = 0$, $u = 1$ on an ellipse with $a/b = 1/2$. For this problem our method is better than that of GILBERT and LINZ [12] when more than 7 terms or so are used. The constancy of error as 12 or more terms are used is probably again due to single precision FORTRAN 77. Adding more terms in the G-function did not change the error significantly. The G-function used here had 11 terms in it.

terms increases and the reason for this unknown. That the errors we get are not the same as those of GILBERT and LINZ [12] for $n = 8$ and 12 can be due to the way we choose the 31 points at which to calculate the error. We do not know how GILBERT and LINZ [12] did this – we distributed the points evenly around the perimeter of the circle starting at angle 0. The G-function was calculated with 11 terms.

The problem $\Delta u - u = 0, u = 1$ on an ellipse with $a/b = 1/2$

The results of our calculation and those of GILBERT and LINZ [12] are seen in Fig. 2. For this problem our method is better than that of [12] when more than 7 terms or so are used. The constancy of error as 12 or more terms are used is probably again due to single precision FORTRAN 77. Adding more terms in the G-function did not change the error significantly. The G-function used here had 11 terms in it.

The solution of a more complicated equation

In the final test we used our program to solve the problem of

$$\Delta u + \frac{8}{(r + r^2)^2} u = 0, \tag{27}$$

where the analytical solution can be shown to be:

$$u(r, \theta) = \tfrac{3}{8} (\sin(2 \arctan(r)) + 5 \sin (6 \arctan(r))) \cos (\theta)$$

we choose the boundary

$$r = 0.25/\sqrt{0.25 \sin (\theta)^2 + \cos (\theta)^2} . \tag{28}$$

As can be seen in Fig. 3 this problem is amenable to increasing the number of terms in the sum over basis functions. For the 6 and 24 term solutions we fitted the approximate solutions at both 32 and 64 points along the boundary to see if this would reduce the error of the fit. As can be seen, the reduction in error is small, so the addition of more fitting points only pays off when the error is small – at 24 terms the reduction is, in this case, equal to an order of magnitude.

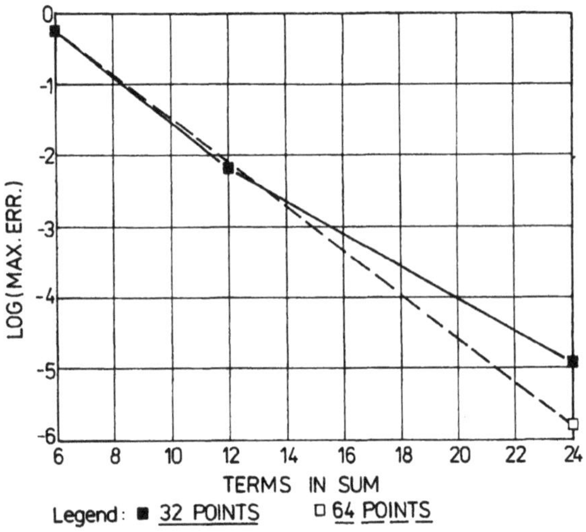

Fig. 3. The analytical solution of $\Delta u + 8u/(1 + r^2)^2 = 0$, can be shown to be: $u(r, \theta) = -(\sin (2 \arctan (r)) + 5 \sin (6 \arctan (r))) \cos (\theta)\ 3/8$. We solved the problem with our method using the boundary

$$r = 0.25/\sqrt{0.25 \sin (\theta)^2 + \cos(\theta)^2} .$$

This problem is amenable to increasing the number of terms in the sum over basis functions. For the 6 and 24 term solutions we fitted the approximate solutions at both 32 and 64 points along the boundary to see if this would reduce the error of the fit. The reduction in error is small, so the addition of more fitting points only pays off when the error is small – at 24 terms the reduction is, in this case, equal to an order of magnitude.

Acknowledgement

The research of P. THEJLL was supported by the NASA (Grant NAG 8-564), by the National Science Foundation (Grant No. AST-85-15747), by the A. J. Cannon Fund and by the Augustinus Foundation in Denmark; that of R. P. GILBERT in part by the National Science Foundation through NSF Grant No. DN 8601318.

References

1 BERGMAN, S: Integral operators in the theory of linear partial differential equations. Ergeb. Math. u. Grenzgeb. 23, Springer Verlag, Heidelberg 1969.
2 BOGEN, R. et al.: MACSYMA-Reference manual, version ten. Symbolics Inc., Cambridge, MA 1984.
3 CARROLL, R. W.: The Bergman-Gilbert operator as a transmutation. C.F. Math. Rep. Acad. Sci. Canada **4** (1982) 5, 267–269.
4 CARROLL, R. W.: Some remarks on the Bergman-Gilbert integral operator. J. Math. pures Appl. **63** (1984), 1–14.

5 COLTON, D. L.; GILBERT, R. P.: On the numerical treatment of partial differential equations by function theoretic methods. In: Proc. Sympos. Numerical Solution of Partial Differential Equations, University of Maryland, May 1970, Academic Press, New York 1971.
6 GAIER, D.: Vorlesungen über Approximationen im Komplexen. Birkhäuser, Stuttgart 1980.
7 GARABEDIAN, P. R.: Partial differential equations. J. Wiley, New York 1964.
8 GILBERT, R. P.; JENSEN, JARL: A computational approach for constructing singular solutions of one-dimensional pseudoparabolic and metaparabolic equations. SIAM J. Sci. Statist. Comput. 3 (1982) 1, 111–125.
9 GILBERT, R. P.: A method of ascent. Bull Amer. Math. Soc. **75** (1969) 6, 1286–1289.
10 GILBERT, R. P.: Function theoretic methods in partial differential equations. Academic Press, New York 1969.
11 GILBERT, R. P.: Constructive methods for elliptic partial differential equations. Lecture Notes in Mathematics 365, Springer-Verlag, Heidelberg 1974.
12 GILBERT, R. P.; LINZ, P.: The numerical solution of some elliptic boundary value problems by integral operator methods. In: Constructive and computational methods in differential integral equations. Lecture Notes in Mathematics 430, Springer-Verlag, Heidelberg 1974.
13 GILBERT, R. P.; WOOD, D. H.: A transmutation approach to underwater sound propagation. Wave Motion **8** (1986) 5, 383–397.
14 HERRERA, I.: Boundary methods. Pitman, Boston 1984.
15 KRZYWOBLOCKI, M. Z.: Bergman's and Gilbert's operators in elasticity, electromagnetism, fluid dynamics, wave mechanics. In: Analytic methods in mathematical physics. Gordon & Breach, New York 1970, pp. 207–247.
16 RUNDELL, W.: The use of integral operators in undetermined coefficient problems for partial differential equations. Appl. Anal. **18** (1984), 309–324.
17 VEKUA, I. N.: New methods for solving elliptic equations. J. Wiley, New York 1967.

Received June 20, 1989

Addresses: Dr. PETER THEJLL, Department of Physics, University of Delaware, Newark, Delaware 19716, USA; Dr. ROBERT P. GILBERT, Professor of Mathematics, Department of Mathematical Sciences, University of Delaware, Newark, Delaware 19716, USA

Appendix A: The main program

```
transmute(): =(
        input(),
        print("input n=", n, "rhs=", rhs, "bdfct=", bdfct),
        gfunction(m),
        print("The g-function has been calculated with" ,m,"terms."),
        saveg(),
        print("The g-function has been saved for future use as lastg"),
        phis(),
        savephi(),
        print("The phis have now been saved for future use as lastphis"),
        overdet(),
        saver()
        );

input(): =(
        answer:read("Do you want to use the old
                input that was saved last time?Y or N"),
        if answer = N then
        (
        m:read("Number of terms in the g-function?"),
        n:read("Number of terms in the
                sum over base functions?"),
        rhs:read("What is the p(", r^2,") function?
                NOTE: Form is nabla(u) + Pu = 0."),
bdfct:read("Give the function u(angle) on the boundary.
                You MUST use 'angle'!"),
        var:read("What was the variable you used above?"),
        radius:read("In plane polar coordinates,
                what is the radial coordinate in terms of the angle coordinate?"),
        bdfct:ev(bdfct),                /* this is needed to incorporate any var dependence */
        save([inputs], n, m, rhs, bdfct, var, radius)
        )
        else
        loadfile(inputs),
        bdfct:ev(bdfct)
        );

gfunction(m): =(
        answer:read("Do you want to use the last g-function calculated? Y or N"),
        if answer = N then
        (
        define(p(r), rhs),
```

```
        v(r): = integrate(r*p(r), r),
        c[1]:v(r),
        for i:1 thru m  do(
                        c[i + 1]:1/(2*i)*integrate((-r*diff(c[i], r, 2)
                        + (2*i - 1)*diff(c[i], r, 1)
                        + r*p(r)*c[i]), r)
                        ),
        define(g(r, t), sum([i]*t^(i - 1), i, 1, m))
        )
        else
                loadfile(lastg),
        print("gfunction =", g(r, t))
        );

phis(): =
(
        answer:read("Do you want to use the last calculated phi's? Y or N"),
        if answer = N then
        for i:1thru n do(
        k:entier(i/2),
        if k = i/2 then
                choice(var): = sin(k*var)
                else
                choice(var): = cos(k*var),
        arg:integrate(sigma^(2*k + 1)*g(r, 1 - sigma^2), sigma),
        mkr:subst(1, sigma, arg)-subst(0, sigma, arg),
        phi[1] (r, var): = r^k*choice(var)*(1 + mkr),
        print("phi[", i, "] = ", phi[i] (r, var))
                )
                else
        loadfile(lastphis),
        for i:1 thru n do
                print("phi[", i, "] = ", phi[i] (r, var))
);


saveg(): = (
        save([lastg], g)
        );

savephi(): =
(
        save([lastphis], phi)
        );

overdet(): =
(
     /*  The problem of fitting n functions to m points, where m is  */
     /*  greater than n is handled here. The case m = 31 is done.  */
     /*    This version of 'overdet' is intended for use on a boundary  */
     /*  like a cardioid where there is a cusp. For THIS version the cusp  */
     /*  is at 180 degrees, and the density of points around the cusp is  */
     /*  increased by a factor of 'factor'  */
/* .................................................................  */
     /*          First open and close the file everything is written  */
     /*  to - since that'll allow us to merely append stuff later:  */
     writefile("erase.me"),
     print("c  This is a comment"),
     closefile("erase.me"),
/* .................................................................  */
factor:read("By what factor is the density of points to be increased?"),
hidensitystart:read("Angle to start hi-density points at?"),
for i:1 thru n do(
    angle:0.0,
    anglecount:0,
for iangle:1 step 1 while angle<=2*%pi do(
    anglestep:2*%pi/31,
    if angle>hidensitystart and angle<(2*%pi-hidensitystart) then
    (
                anglestep:2*%pi/31/factor
    ),
    r:ev(radius, numer),
/* a term is put in the matrix of the overdetermined system:  */
```

```
    writeoverdetmatrix(iangle, 1, ev(phi[i] (r, angle), numer)),
    angle:angle + anglestep,
    anglecount:anglecount + 1
                                ) /* end of iangle loop */
                ), /* end of i loop */
/* write the b terms: */
    angle:0.0,
    anglecount:0,
for iangle:1 step 1 while angle< =2*%pi do(
    anglestep:2*%pi/31,
    if angle>hidensitystart and angle<(2*%pi-hidensitystart) then
    (
                anglestep:2*%pi/31/factor
    ),
    r:ev(radius, numer),
    writeoverdetmatrix(iangle, 2, ev(bdfct, numer)),
    angle:angle + anglestep,
    anglecount:anglecount + 1
                                ), /* end of iangle loop #2 */
/* Now turn that file into a FORTRAN program: */
    dummy = 0, /* to LLSQF etc. */
    writeoverdetmatrix(iangle, 3, dummy),
/* Now that A.x = B is defined solve it using IMSL-routine LLSQF: */
    solveoverdetermined(),
/* The solution is known as the array a[1],.........., a[n] */
/* so go ahead and use it to make up the approximate solution */
/* and find the greatest error on the boundary at the 31 points */
/* in the sense of Linz & Gilbert [12] */
    makesolution(),
    outputter()
);


makesolution(): =
(
    /* Do the sum a[i]*phi[i] (r, var) which will be the approximate */
    /* solution to our problem. */
        ufct(radial, var): = sum(a[i]*phi[i] (radial, var), i, 1, n)
);


writeoverdetmatrix(index, switch, argument): =
    (
    /* Switch = 1 means argument is from the matrix A in A.x = B */
    /* Switch = 2 means argument is from the vector B above */
    /* Switch = 3 means turn file into propper FORTRAN */

    if switch = 1 then
      (
      appendfile("erase.me"),
      print("       a(", index,",", i, ") = ", ev(argument, numer)),
      closefile("erase.me")
),
    if switch = 2 then
(
      appendfile("erase.me"),
      print("       b(", ev(index, numer),") = ", float(ev(argument, numer))),
      closefile("erase.me")
),
    if switch = 3 then
(
      writefile("front.mac"),
print("       double precision a(", anglecount, ", ", n, "), b(", anglecount, "), "),
print("       & tol, x(", n, "), h(", n, ")"),
print("       integer ia, m, n, kbasis, ip(", n, "), ier"),
print("       open(9, file =", " 'result')"),
      closefile("front.mac"),
      exec("cat front.mac erase.me > hold.for"),        /* UNIX commands */
      appendfile("hold.for"),
print("       ia = ", anglecount),
print("       m = ", anglecount),
print("       n = ", n),
print("       tol = 0.0"),
```

```
print("        kbasis = ", n),
print("        call llsqf (a, ia, m, n, b, tol, kbasis, x, h, ip, ier)"),
print("        do 1 i = 1, ", n),
        closefile("hold.for"),
        exec("cat hold.for endbit.f > hold")                /* UNIX commands */
        exec("grep -v 'done' hold > hold.f"),
        exec("rm hold.for hold")
)
);


solveoverdetermined() :=
(
        /*   Now compile the file hold.f and run it   */
             exec("f77 hold.f -limsl"),   /* UNIX commands */
             exec("a.out"),
             batch(result)
);


outputter() :=
(
fpprec:10,
writefile("out1"),
fortran(radi = subst(vinkel, var, radius)),
for i:1 thru n do(
        fortran(term(i) = bfloat(a[1]*phi[i] (radi, vinkel)))
                  ),
fortran(f = abs(bdfct-sum(term(i), i, 1, n))),
closefile("out1"),
exec("fgrep -v 'Warning' out1 > out"),
exec("rm out1")
);


saver() :=
(  /*  Here the approximate solution ufct(rad, ang) is saved in a
       file. If you want to avoid overwriting this file in later runs
       you must rename the file 'appsol' using your operating system   */


       save([appsol], ufct(rad, ang))
);
```

ZAMM · Z. angew. Math. Mech. **70** (1990) 11, 489–500 Akademie-Verlag Berlin

PANAYOTOUNAKOS, D. E.; THEOCARIS, P. S.

On the Decoupling and the Solutions of the Euler Dynamic Equations Governing the Motion of a Gyroscope

Es werden vorgestellt die Entkopplung und die Lösungen der drei stark nichtlinearen gewöhnlichen dynamischen Differentialgleichungen, die die Bewegung eines beliebigen starren Körpers beschreiben, der um einen festen Punkt rotieren kann (Kreisel). Die entwickelte Theorie basiert auf der Annahme, daß die momentane Winkelgeschwindigkeit und die Momentresultanten glatte, willkürliche Funktionen der Zeit sind. Es ergeben sich aus der Rechnung unter einigen allgemeinen Bedingungen analytische Lösungen der gewonnenen Differentialgleichungen in Übereinstimmung mit dem physikalischen Problem. Schließlich wird eine theoretische Anwendung besprochen, die die dynamische Reaktion eines einachsigen symmetrischen Kreisels betrifft, der einer beliebigen äußeren Belastung ausgesetzt ist.

The decoupling and the solutions of the three strongly nonlinear ordinary dynamic differential equations, governing the motion of an arbitrary rigid body, free to rotate about a fixed point (gyro), are presented. The theory developed is based on the assumption that the instantaneous angular velocity and the moment-components are arbitrary smooth functions of the time. By a quantitative analysis, analytical solutions of the resulting differential equations were obtained under some general conditions in accordance with the physical problem. Finally, a theoretical application is investigated concerning the dynamic response of a one-axis symmetrical gyroscope subjected to an arbitrary external loading.

Представляется развязывание и решение трех строго нелинейных обыкновенных динамических дифференциальных уравнений управляющих, движение любого жесткого тела, которое может вращать вокруг неподвижной точки (волчока). Развитая теория основывается на предположение, что и мгновенная угловая скорость и результанты момента являются любюми гладкими функциями от времени. При помощи количественного анализа получаются аналитические решения проистекающих дифференциальных уравнений при некоторых общих условиях являющих в соответствии с физической задачей. Наконец изучается теоретичное применение касающееся динамический ответ одноосевого симметричного гироскопа при условии любой внешной нагрузки.

1. Introduction

There is a large number of studies on the motion of an arbitrary rigid body free to rotate about a fixed point. A detailed bibliography on this subject may be found in references [1] to [7]. The three dynamic and three kinematic Euler's equations, governing this problem, were mainly applied to the theory of selfexcited gyros (cf. [1] to [13]) and they were numerically or analytically solved for special cases of motion.

A general theory on the possibility of decoupling and solving Euler's equations for selfexcited rigid-body motions was developed by MAGNUS [1] and WITTENBURG [2]. Also, the classical studies by GRAMMEL [3, 4], dealing with the analytical solutions of the beforementioned equations, should be mentioned. In addition, more recent publications on the investigation of gyros are those by IL'CHANICOV and TERESHIN [8] and by SMOL'NIKOV [9]. In the first one [8], an exact analytical integration of a nonlinear mechanical system was constructed on the basis of the behavior of a one-axis gyroscope, mounted on a uniformly rotating base, when the suspension ring is statically unbalanced. In the second paper ([9]), SMOL'NIKOV succeeded in integrating the Euler dynamic equations assuming that the kinetic moment-vector keeps its magnitude unchanged, although its direction in space alters. FOX and BURDESS [10] investigated the dynamics of a gyroscope, supported on a heavy elastic suspension, using a mathematical model which allowed the gyroscope to be treated as a two-degree-of-freedom rigid body on a light suspension. Moreover, SMOL'NIKOV's [11] as well as ORMANDY's and MAUNDER's contributions [12] to the dynamics of a rigid body should be mentioned. Finally, PANAYOTOUNAKOS and THEOCARIS [13] succeeded in giving analytical solutions of the Euler kinematic equations of a selfexcited gyro through general conditions in accordance with the physical problem.

This paper deals with the problem of decoupling and subsequently establishing analytical solutions, under general conditions, of the three Euler dynamic equations governing the motion of a gyro. The theory developed is based on the assumptions that the instantaneous angular velocity and the moment components are arbitrary smooth functions of the time. Also, the referred coordinate system coincides with the principal axes of inertia for the fixed point, moving with the body.

In the second section we succeeded in presenting a general technique for the decoupling of the Euler dynamic equations. This decoupling technique leads to a third-order strongly nonlinear ordinary differential equation (of variable coefficients) with respect to one of the angular velocity-components; or, to two differential equations of lower order by means of which estimation of the analytical solutions seems to be possible.

In the third section, by a quantitative analysis concerning the previous two equations, we establish analytical solutions under general conditions in accordance with the physical problem. The developed mathematical procedure and, mainly, the functional transformations introduced in order to obtain the beforementioned analytical solutions dispense the researcher from the necessity of running long and costly numerical procedures, which not only lack completeness and accuracy, but also need large memory space and coding. Finally, in the fourth section an interesting application is examinend for a one-axis symmetrical gyro subjected to a generic loading.

2. The decoupling procedure

Consider the motion of an arbitrary rigid body, free to rotate about a fixed point 0 (gyro). Let us denote by $0x_1y_1z_1$ the space-fixed coordinate system, while $0xyz$ be the body-fixed coordinate system. The state motion is a rotation with an instantaneous angular velocity $\boldsymbol{\omega} = (\omega_x, \omega_y, \omega_z)$ with respect to the $0xyz$-system. Let J_x, J_y, J_z denote the principal moments of inertia of the body, corresponding to the x,y,z-axes, respectively, with $J_x < J_y < J_z$. Then the body is asymmetric and so the ellipsoid of inertia does not have rotational symmetry. In the absence of friction, the external forces acting on the gyro reduce to a simple force $\boldsymbol{R}$ and a couple $\boldsymbol{M}$ at 0. Consequently, the vector $\boldsymbol{M}$ is known and the motion of the body can be obtained by the angular momentum theorem.

On the basis of the previous assumptions we may derive the well-known Euler dynamic equations with respect to the principal axes of inertia x, y, z as follows:

$$J_x\dot{\omega}_x - a_1\omega_y\omega_z = M_x, \qquad J_y\dot{\omega}_y - a_2\omega_x\omega_z = M_y, \qquad J_z\dot{\omega}_z - a_3\omega_x\omega_y = M_z, \qquad (2.1\text{a}, 1\text{b}, 1\text{c})$$

where

$$a_1 = J_y - J_z, \qquad a_2 = J_z - J_x, \qquad a_3 = J_x - J_y, \qquad (2.2)$$

and a dot means differentiation with respect to time t. If the moment-components M_x, M_y, M_z are smooth functions only of time, then the Euler dynamic equations (2.1) consist of a set of three strongly nonlinear first-order ordinary differential equations, the integration of which constitutes a complex mathematical problem. Several investigations have been developed, mainly on selfexcited rigid-body motions, and various analytical, approximate and numerical methods are usually employed to solve specific problems (mainly for symmetrical gyros, see [1] to [7]).

The aim of this paper is, at first, to present a general technique for the decoupling of the dynamic system (2.1) and formulate an equivalent higher-order nonlinear differential equation with respect to one of the components ω_x, ω_y, ω_z. In the sequel, by a quantitative analysis, we shall try to estimate analytical solutions of the last equation by way of general conditions in accordance with the physical problem.

The introduced general procedure of decoupling is the following. We multiply equations (2.1) by the nonzero and smooth functions ω_x, ω_y, ω_z, respectively, solve the resulting new equations with respect to ω_x, ω_y, ω_z and combine the results. Then we deduce

$$\omega_x(\dot{\omega}_x - f_1) = \lambda_1\omega_y(\dot{\omega}_y - f_2), \qquad \omega_y(\dot{\omega}_y - f_2) = \lambda_2\omega_z(\dot{\omega}_z - f_3), \qquad (2.3), (2.4)$$

in which

$$\lambda_1 = A_1/A_2, \qquad \lambda_2 = A_2/A_3, \qquad A_1 = (J_y - J_z)/J_x, \qquad A_2 = (J_z - J_x)/J_y,$$

$$A_3 = (J_x - J_y)/J_z, \qquad f_1(t) = M_x(t)/J_x, \qquad f_2(t) = M_y(t)/J_y, \qquad f_3(t) = M_z(t)/J_z. \qquad (2.5)$$

On the other hand, by means of (2.1a) and (2.1c) one gets, respectively:

$$\dot{\omega}_x = f_1 + A_1\omega_y\omega_z, \qquad \omega_x = (\dot{\omega}_z - f_3)/A_3\omega_y. \qquad (2.6), (2.7)$$

Differentiating the last equation once with respect to time and combining the second member of the resulting new equation together with the corresponding member of (2.6), we obtain

$$\omega_y\ddot{\omega}_z - \dot{\omega}_y\dot{\omega}_z - C_1\omega_y^3\omega_z = h_1, \qquad (2.8)$$

where

$$C_1 = A_1A_3, \qquad h_1 = \omega_y^2 g_1 + \omega_y\dot{f}_3 - \dot{\omega}_y f_3, \qquad g_1 = A_3 f_1. \qquad (2.9)$$

Simultaneously, differentiation of (2.4) with respect to time results in

$$\lambda_2\omega_z\ddot{\omega}_z + \lambda_2\dot{\omega}_z^2 - \lambda_2 f_3\dot{\omega}_z - \lambda_2\dot{f}_3\omega_z = h_2, \qquad (2.10)$$

in which

$$h_2 = \omega_y\ddot{\omega}_y + \dot{\omega}y^2 - f_2\dot{\omega}_y - \dot{f}_2\omega_y. \qquad (2.11)$$

From now on, by using equations (2.4), (2.8) and (2.10), we shall try to eliminate the terms ω_z, $\dot{\omega}_z$ and $\ddot{\omega}_z$. For this purpose, introducing the ω_z-component from (2.4) into (2.8), after some algebra, we deduce

$$(\omega_y\ddot{\omega}_z - \dot{\omega}_y\dot{\omega}_z - h_1)(\dot{\omega}_z - f_3) = \omega_y^3 h_3, \qquad (2.12)$$

where

$$h_3 = C_1 g_2/\lambda_2, \qquad g_2 = \omega_y(\dot{\omega}_y - f_2). \qquad (2.13)$$

Substituting now the term $\lambda_2\omega_z$ given in (2.10) by the corresponding expression included in (2.4), we formulate the equation

$$g_2\ddot{\omega}_z + [\lambda_2\dot{\omega}_z(\dot{\omega}_z - f_3) - h_2](\dot{\omega}_z - f_3) = g_2\dot{f}_3 \,. \tag{2.14}$$

Solving this equation for $\ddot{\omega}_z$ and using (2.12), we have succeeded in deducing the following quadric equation for $\dot{\omega}_z$:

$$\dot{\omega}_z^4 + a\dot{\omega}_z^3 + b\dot{\omega}_z^2 + c\dot{\omega}_z + d = 0\,, \tag{2.15}$$

in which

$$a = 3f_3\,, \qquad b = 3f_3^2 - (h_2/\lambda_2) + (g_2\dot{\omega}_y/\lambda_2\omega_y)\,,$$

$$c = f_3\{-f_3 + (2h_2/\lambda_2) - (g_1\dot{\omega}_y/\lambda_2\omega_y)\} - g_2[\dot{f}_3 - (h_1/\omega_y)]/\lambda_2\,, \tag{2.16}$$

$$d = \{f_3\{-f_3h_2 + g_2[f_3^2 - (h_1/\omega_y)]\}/\lambda_2\} + g_2\omega_y^2h_3/\lambda_2\,.$$

One may observe here, that the coefficients of (2.15) are smooth functions of the ω_y-component and its derivatives, as well as of the moment-components and their derivatives. Consequently, supposing that this equation is solvable for $\dot{\omega}_z$, one my write

$$\dot{\omega}_z = R(\omega_y, \dot{\omega}_y, \ddot{\omega}_y; f_1, f_2, f_3; \dot{f}_1, \dot{f}_2, \dot{f}_3), \tag{2.17}$$

where R respresents a real root of (2.15). On the other hand, by way of (2.4), one obtains

$$\dot{\omega}_z = [\omega_y(\dot{\omega}_y - f_2)/\lambda_2(\dot{\omega}_z - f_3)]^{\cdot}\,.$$

From the above analysis it is clear, that the introduction of the expression (2.17) and its derivatives into the last equation constructs a strongly nonlinear third-order ordinary differential equation (of variable coefficients) with respect to the ω_y-component, equivalent to the initial system (2.1). Furthermore, if the general integral of this equation is known, then the remaining ω_z- and ω_x-components are evaluated by means of (2.15) and (2.7), respectively. But, since the coefficients of the quadric equation (2.15) are functions of the ω_y-component and its derivatives, the estimate of the roots of this equation is a rather complicated problem, and consequently the formulation and especially the solution of the above third order differential equation seems to be impracticable. However, as we shall see in the following, the formulation of an additional differential equation with respect to the ω_y-component will prove effective for the estimate of analytical solutions in combination with (2.15). In fact, by means of (2.4), we have

$$\omega_y\dot{P}_y + \dot{\omega}_yP_y = \lambda_2(\dot{\omega}_zq_z + \omega_z\dot{q}_z)\,, \tag{2.18}$$

where

$$P_y = \dot{\omega}_y - f_2\,, \qquad q_z = \dot{\omega}_z - f_3\,. \tag{2.19}$$

Solving (2.4) for ω_z and substituting the result into (2.19), we obtain

$$\omega_y\dot{P}_y + \dot{\omega}_yP_y = \lambda_2[\dot{\omega}_zq_z + (\dot{q}_z\omega_yP_y/\lambda_2q_z)]\,,$$

which, by way of (2.19) and after some algebra, results in

$$(\omega_y^2)^{\cdot\cdot} - \sigma_1(\omega_y^2)^{\cdot} - 2f_2\dot{\omega}_y - \sigma_2\omega_y = \sigma_3\,, \tag{2.20}$$

in which

$$\sigma_1 = \dot{q}_z/q_z\,, \qquad \sigma_2(\dot{f}_2 - f_2\sigma_1)\,, \qquad \sigma_3 = 2\lambda_2\dot{\omega}_zq_z\,. \tag{2.21}$$

The nonlinear second-order differential equation (2.20) concerns the ω_y-component and it is remarkable that its coefficients are functions of the ω_z-component and its derivatives, as well as of the moment-components and their derivatives. Also, from the beforementioned decoupling it is clear that both (2.15) and (2.20) are equivalent to the already constructed third-order nonlinear differential equation with respect to the ω_y-component.

For a complete quadric equation it is well-know that, if we put

$$\dot{\omega}_z = \psi_z - (a/4)\,, \tag{2.22}$$

in (2.15), we can deduce a new form:

$$\psi_z^4 + \alpha\psi_z^2 + \beta\psi_z + \gamma = 0\,, \tag{2.23}$$

in which

$$\left.\begin{aligned} &\alpha = -(3a^2/8) + b\,, \qquad \beta = \{a[(a^2/4) - b]/2\} + c\,, \\ &\gamma = \{a[-(3a^3/64) + (ab/4) - c]/4\} + d\,. \end{aligned}\right\} \tag{2.24}$$

Also it is well-known, that the four roots of (2.23) can be evaluated by the formulae

$$\left.\begin{aligned} 2\psi_{z1} &= (z_1)^{1/2} + (z_2)^{1/2} + (z_3)^{1/2}, & 2\psi_{z2} &= (z_1)^{1/2} - (z_2)^{1/2} - (z_3)^{1/2}, \\ 2\psi_{z3} &= -(z_1)^{1/2} + (z_2)^{1/2} - (z_3)^{1/2}, & 2\psi_{z4} &= -(z_1)^{1/2} - (z_2)^{1/2} + (z_3)^{1/2}, \end{aligned}\right\} \tag{2.25}$$

where z_1, z_2, z_3, z_4 denote the roots of the cubic equation

$$z^3 + 2\alpha z^2 + (\alpha^2 - 4\gamma) z - \beta^2 = 0. \tag{2.26}$$

Finally, by the substitution

$$z = x - (2\alpha/3) \tag{2.27}$$

equation (2.26) is reduced to the Cardan form

$$x^3 + Ax + B = 0, \tag{2.28}$$

in which

$$A = -(\alpha^2/3) - 4\gamma, \qquad B = -(2\alpha^3/27) + (8\alpha\gamma/3) - \beta^2. \tag{2.29}$$

The discriminant

$$D = -27B^2 - 4A^3$$

of the Cardan equation (2.28) coincides with the corresponding discriminant of the quadric equation (2.23) or (2.15), and it can be evaluated as

$$\bar{D} = 16\alpha^4\gamma - 4\alpha^3\beta^2 - 128\alpha^2\gamma^2 + 144\alpha\gamma\beta^2 + 256\gamma^3 - 27\beta^4.$$

Now, we distinguish the following four possible cases:

i) if $D < 0$, then two of the roots of (2.15) are real and two are complex conjugate;

ii) if $D > 0$ and $\alpha < 0$; $\alpha^2 - 4\gamma > 0$, then the four roots of (2.15) are real;

iii) if $D > 0$ and $\alpha \geqq 0$ or $\alpha^2 - 4\gamma < 0$, then the four roots of (2.15) are complex conjugate;

iv) if $D = 0$, then two of the roots of (2.15) are real and equal, while the other two are real or complex conjugate when the one root of (2.26) is positive or negative, respectively;

v) if $\alpha^2 + 12\gamma = 0$; $8\alpha^3 + 27\beta^2 = 0$, then three of the roots of (2.15) are equal, and

vi) if $\beta = 0$; $\alpha^2 - 4\gamma = 0$, then (2.15) has two double roots. If $\alpha > 0$, these roots are real, and, if $\alpha < 0$ they are complex conjugate.

From the above investigation for the roots of the quadric equation (2.15) we conclude, in principle, that all the real roots (positive or negative) of this equation are accepted as solutions of the problem, since the resultant $\dot{\omega}_z$ can take positive or negative values. On the other hand, the physical problem under consideration would require a unique solution. Consequently, according to the previous developments, one may conclude that

$$z_2 = z_3 = 0, \qquad z_1 = \text{positive}, \qquad \psi_{z1} = \psi_{z2} = (z_1)^{1/2}/2, \qquad \psi_{z3} = \psi_{z4} = -(z_1)^{1/2}/2,$$

a fact that requires the validity of the following relations connecting the coefficients of the quadric equation (2.23)

$$\beta = 0, \qquad \alpha^2 - 4\gamma = 0, \qquad \alpha > 0. \tag{2.30}$$

3. Analytical solutions of the equations governing the motion of the gyroscope

We shall try to estimate analytical solutions of equations (2.15) and (2.20) under general conditions in accordance with the physical problem. These conditions concern the moment-components depending on the external loading.

Consider, in principle, the motion of a gyro in which one moment-component, the M_z-component, say, is missing, while the other two are smooth functions of time. Then, from (2.5) we conclude that the function f_3 is equal to zero, while f_1 and f_2 are smooth functions of time. Consequently, the coefficients (2.16), concerning the quadric equation (2.15), become:

$$a = 0, \qquad b = -[h_2 - (g_2\dot{\omega}_y/\omega_y)]/\lambda_2, \qquad c = g_2 h_1/\lambda_2\omega_y, \qquad d = g_2\omega_y^2 h_3/\lambda_2. \tag{3.1}$$

Using relations (2.24), which connect the coefficients of both quadric equations (2.23) and (2.15), as well as (3.1), we are led to

$$\alpha = b > 0, \qquad \beta = c = 0, \qquad \gamma = d. \tag{3.2}$$

The combination of (3.2) and (2.30) furnishes

$$b^2 - 4d = 0, \qquad b > 0, \qquad c = 0. \tag{3.3); (3.4}$$

Taking into account the third equation of (2.16), as well as that $f_3 = 0$, we deduce from (3.4) the relation

$$g_2 h_1/\lambda_2 \omega_y = 0\,, \tag{3.5}$$

in which

$$g_2 = \omega_y(\dot{\omega}_y - f_2)\,, \qquad h_1 = \omega_y^2 g_1\,, \qquad g_1 = A_3 f_1\,. \tag{3.6}$$

Consequently, by way of (3.5) and for $\omega_y \neq 0$, we estimate the following three solutions:

i) $g_2 = 0, \quad h_1 = 0 \Rightarrow \omega_y(\dot{\omega}_y - f_2) = 0, \quad f_1 = 0;$

ii) $g_2 = 0 \Rightarrow \omega_y(\dot{\omega}_y - f_3) = 0;$

iii) $h_1 = 0 \Rightarrow f_1 = 0.$

We shall investigate now separately the three beforementioned solutions.

Solutions (i) and (ii).

These two solutions are trivial. In fact, since $\omega_y \neq 0$, the common equation $g_2 = 0$ gives

$$\omega_y(\dot{\omega}_y - f_2) = 0\,, \tag{3.7}$$

from which we obtain:

$$\omega_y = \int f_2 \,\mathrm{d}t + c_1\,, \tag{3.8}$$

where c_1 is an integration constant.

On the other hand, the first relation of (3.3) by means of (3.1) results in

$$h_2 = 0\,,$$

which, according to (2.11), furnishes

$$\omega_y(\dot{\omega}_y - f_2) = c\,,$$

where c is a constant.

But, since (3.7) is valid, the constant c must be equal to zero. (2.4) furnishes

$$\omega_z = c_2 = \text{const.}$$

and (2.3)

$$\omega_x = \int f_1 \,\mathrm{d}t + c_3\,, \quad \text{if} \quad f_1 \neq 0\,, \qquad \omega_x = c_3\,, \quad \text{if} \quad f_1 = 0\,, \tag{3.9}$$

where c_3 is a new constant.

Solution (iii):

Here, since $f_1 = 0$, we conclude that the moment-component M_x is equal to zero; so, in this case, only one moment-component, viz . M_y, exists, which is an arbitrary smooth function of time. We thus have

$$M_x = M_z = 0\,; \qquad h_1 = f_1 = g_1 = 0\,. \tag{3.10}$$

The first relation of (3.3), by way of (2.16), results in

$$\left(g_2 \frac{\dot{\omega}_y}{\lambda_2 \omega_y} - \frac{h_2}{\lambda_2}\right)^2 = 4 \frac{g_2 \omega_y^2 h_3}{\lambda_2}\,, \tag{3.11}$$

which, since $b > 0$, furnishes

$$g_2 \frac{\dot{\omega}_y}{\lambda_2 \omega_y} - \frac{h_2}{\lambda_2} = 2 \frac{g_2^{1/2} \omega_y h_3^{1/2}}{\lambda_2^{1/2}}\,. \tag{3.12}$$

But, according to the first equation of (2.13), one gets

$$2 g_2^{1/2} \omega_y h_3^{1/2} / \lambda_2^{1/2} = 2 C_1^{1/2} g_2 \omega_y / \lambda_2\,.$$

Using the last expression, as well as (2.8) and the second equation of (2.13), we transform (3.12) into the form

$$\dot{\omega}_y(\dot{\omega}_y - f_2) - (\omega_y \dot{\omega}_y)^{\cdot} + (\omega_y f_2)^{\cdot} = 2 \sqrt{C_1}\, \omega_y^2 (\dot{\omega}_y - f_2)\,. \tag{3.13}$$

Since $\omega_y \neq 0$ and after some algebra this equation is reduced to the form

$$(\dot{\omega}_y - f_2)^{\cdot} + 2 \sqrt{C_1}\, \omega_y (\dot{\omega}_y - f_2) = 0\,. \tag{3.14}$$

Equation (2.14) constructs the second-order nonlinear ordinary differential equation of variable coefficients with respect to the ω_y-component, which, in the present case, is equivalent to the general equation (2.20). From now on, we shall prove, by suitable functional transformations, that this equation can be reduced to an integrable form.

At firstly, by the substitution:

$$\omega_y = u + P_2, \qquad P_2(t) = \int f_2 \, dt \tag{3.15}$$

we obtain

$$\ddot{u} + 2\sqrt{C_1}\,(u + P_2)\,\dot{u} = 0 \tag{3.16}$$

which is equivalent to (3.14). We introduce now the functional transformation

$$u(t) = n(\xi), \qquad \xi = \int e^{-2\sqrt{C_1}\int P_2 dt} \, dt, \tag{3.17}$$

which, by way of the chain rule, furnishes

$$\dot{u} = (du/dt) = (du/d\xi)(d\xi/dt) = e^{-2\sqrt{C_1}\int P_2 dt}\,\overset{*}{\eta}, \qquad \ddot{u} = e^{-4\sqrt{C_1}\int P_2 dt}\,\overset{**}{\eta} - 2\sqrt{C_1}\,P_2\,e^{-2\sqrt{C_1}\int P_2 dt}\,\overset{*}{\eta}, \tag{3.18}$$

where stars here indicate differentiation with respect to ξ. Inserting (3.18) into (3.16), and after some algebra, we deduce the equation

$$\overset{**}{\eta} + (2\sqrt{C}/\dot{\xi})\,\eta\overset{*}{\eta} = 0, \tag{3.19}$$

where

$$1/\dot{\xi} = dt/d\xi = e^{2\sqrt{C_1}\int P_2 dt} = g(t). \tag{3.20}$$

Since P_2 is a smooth function of time, g is also a smooth function, a fact that permits us to write

$$\overset{*}{\eta} = (dn/d\xi) = (dn/dg)\cdot(dg/d\xi) = n'\overset{*}{g}, \qquad \overset{**}{\eta} = (n'\overset{*}{g})'\,\overset{*}{g}, \tag{3.21}$$

where prime, here, denotes differentiation with respect to g. Then (3.19) is transformed to

$$(n'\overset{*}{g})' + 2\sqrt{C_1}\,(gn')\,\eta = 0. \tag{3.22}$$

But, by means of (3.20) we have

$$\overset{*}{g} = dg/d\xi = (dg/dt)(dt/d\xi) = \dot{g}/\dot{\xi} = 2\sqrt{C_1}\,P_2 g^2$$

and therefore (3.22) is reduced to

$$(P_2 g^2 n')' + gnn' = 0. \tag{3.23}$$

Supposing, without loss of generality, that the external loading is a harmonic function of time, i.e. that

$$f_2(t) = \frac{M_2(t)}{J_y} = \frac{\overline{M}_2}{J_y}\,e^{i\overset{*}{\omega}t}, \qquad i = (-1)^{1/2}, \tag{3.24}$$

where $\overline{M}_2$ is a constant and $\overset{*}{\omega}$ denotes the eigenfrequency of the loading, we obtain:

$$P_2(t) = \int f_2 \, dt = -\frac{i}{\overset{*}{\omega}} f_2, \qquad \dot{\xi} = \frac{1}{g(t)} = e^{(2i\sqrt{C_1}/\overset{*}{\omega})f_2},$$

$$\ln g = \ln(1/\dot{\xi}) = -(2i\sqrt{C_1}/\omega^*)\,f_2 = 2\sqrt{C_1}\,P_2. \tag{3.25}$$

Thus (3.23) results in

$$(g^2 \ln g n')' + 2\sqrt{C_1}\,gnn' = 0. \tag{3.26}$$

A new functional transformation is now the following:

$$n(g) = y(x), \qquad x = 1/g, \tag{3.27}$$

which, since $n' = du/dg = dy/dg = (dy/dx)(dx/dg) = -\dot{y}/g^2$ where dot here means differentiation with respect to x, reduces (3.26) to

$$x(\ln x\dot{y})^{\dot{}} + 2\sqrt{C_1}\,y\dot{y} = 0. \tag{3.28}$$

Rewriting the last equation as

$$x \ln x \ddot{y} + (1 + 2\sqrt{C_1}\, y)\, \dot{y} = 0\,, \tag{3.29}$$

and using the substitution

$$z(x) = 1 + 2\sqrt{C_1}\, y(x)\,, \tag{3.30}$$

we obtain, for $x \ln x \neq 0$,

$$z + z\dot{z}/x \ln x = 0\,. \tag{3.31}$$

Because of $z\dot{z}/x \ln x = \frac{1}{2}\,[z^2(\ln(\ln x))]^{\cdot}$, the last equation can be written as

$$[2\dot{z} + z^2(\ln(\ln x))]^{\cdot} = 0\,,$$

from the integration of which we deduce

$$2\dot{z}(x) + z^2(\ln(\ln x)) = C_2\,, \tag{3.32}$$

where C_2 is a new constant of integration.

For a further integration of (3.32) we put

$$\omega = \ln(\ln x) \Rightarrow x = \exp(\exp(\omega))\,, \qquad \mathrm{d}x = \mathrm{e}^{\omega}\, \mathrm{e}^{\mathrm{e}^{\omega}}\, \mathrm{d}\omega \tag{3.33}$$

and, based on (3.32), we guess that

$$2\, \mathrm{d}z/(C_2 - z^2) = \mathrm{e}^{\omega}\, \mathrm{e}^{\mathrm{e}^{\omega}}\, \mathrm{d}\omega\,, \qquad C_2 > z^2\,,$$

from the integration of which we obtain:

$$\frac{1}{i\sqrt{C_2}} \ln \frac{C_2 + i\sqrt{C_2}\, z}{C_2 - i\sqrt{C_2}\, z} = \mathrm{e}^{\mathrm{e}^{\omega}} + C_3 = x + C_3\,, \tag{3.34}$$

where C_3 is a new constant of integration.

By means of relations (3.30) and (3.27), equation (3.34) is written in the from

$$\ln \frac{C_2 + i\sqrt{C_2}\,(1 + 2\sqrt{C_1}\, n)}{C_2 - i\sqrt{C_2}\,(1 + 2\sqrt{C_1}\, n)} = i\sqrt{C_2}\left(\frac{1}{g} + C_3\right),$$

which, by way of (3.25) and (3.17), results in

$$\ln \frac{C_2 + i\sqrt{C_2}\,(1 + 2\sqrt{C_1}\, n)}{C_2 - i\sqrt{C_2}\,(1 + 2\sqrt{C_1}\, n)} = i\sqrt{C_2}\left(\mathrm{e}^{i\frac{2\sqrt{C_1}}{\overset{*}{\omega}} f_2} + C_3\right). \tag{3.35}$$

Since $f_2 = \overline{M}_2\, \mathrm{e}^{i\overset{*}{\omega}t}/J_y$, equation (3.35) can be expressed as

$$\ln \frac{C_2 + i\sqrt{C_2}\,(1 + 2\sqrt{C_1}\, u)}{C_2 - i\sqrt{C_2}\,(1 + 2\sqrt{C_1}\, u)} = i\sqrt{C_2}\,[\cos(\lambda\, \mathrm{e}^{i\overset{*}{\omega}t}) + i \sin(\lambda\, \mathrm{e}^{i\overset{*}{\omega}t}) + C_3] \tag{3.36}$$

where

$$\lambda = 2\sqrt{C_1}\, \overline{M}_2/J_y\overset{*}{\omega}\,. \tag{3.37}$$

Making use of the well-known formulae

$$\mathrm{e}^{ix} = \cos x + i \sin x\,, \qquad \sin(ix) = i\, \mathrm{sh}\, x\,, \qquad \cos(ix) = \mathrm{ch}\, x\,,$$

(3.36) is reduced to

$$\frac{C_2 + i\sqrt{C_2}\,(1 + 2\sqrt{C_1}\, u)}{C_2 - i\sqrt{C_2}\,(1 + 2\sqrt{C_1}\, u)} = \mathrm{e}^{\omega_1}(\cos \omega_2 + i \sin \omega_2)\,, \tag{3.38}$$

in which

$$\begin{gathered} \omega_1 = \sqrt{C_2}\, \sin g_1(\mathrm{sh}\, g_2 - \mathrm{ch}\, g_2)\,, \qquad \omega_2 = \sqrt{C_2}\,[\cos g_1(\mathrm{ch}\, g_2 - \mathrm{sh}\, g_2) + C_3]\,; \\ g_1 = \lambda \cos \overset{*}{\omega}t\,, \qquad g_2 = \lambda \sin \overset{*}{\omega}t\,. \end{gathered} \tag{3.39}$$

From the initial Euler dynamic equations (2.1) it is clear that, if their right-hand sides are complex functions of time, then their left-hand sides are also complex functions of time. Consequently, u given by (3.15) is a complex function of time, i. e. one may write

$$u = u_1 + iu_2 .$$

The physical meaning of this hypothesis is that the real part of u, or of the ω_y-component, satisfies the cosine form of the external harmonic loading, while the imaginary part satisfies the corresponding sine form. So, (3.38) is equivalent to the equation

$$\frac{(C_2 + 2\sqrt{C_1C_2}\, u_2) + i\sqrt{C_2}\,(1 + 2\sqrt{C_1}\, u_1)}{(C_2 + 2\sqrt{C_1C_2}\, u_2) - i\sqrt{C_2}\,(1 + 2\sqrt{C_1}\, u_1)} = e^{\omega_1}\cos\omega_2 + i\, e^{\omega_1}\sin\omega_2 , \tag{3.40}$$

from which we deduce

$$\begin{aligned} (C_2 - 2\sqrt{C_1C_2}\, u_2) &= (C_2 + 2\sqrt{C_1C_2}\, u_2)\, e^{\omega_1}\cos\omega_2 + \sqrt{C_2}\,(1 + 2\sqrt{C_1}\, u_1)\, e^{\omega_1}\sin\omega_2 , \\ \sqrt{C_2}\,(1 + 2\sqrt{C_1}\, u_1) &= (C_2 + 2\sqrt{C_1C_2}\, u_1)\, e^{\omega_1}\sin\omega_2 - \sqrt{C_2}\,(1 + 2\sqrt{C_1}\, u_1)\, e^{\omega_1}\cos\omega_2 . \end{aligned} \tag{3.41}$$

Solving the last system for u_1, u_2, we deduce

$$u_1 = \frac{1}{2\sqrt{C_1}}\left[\frac{2\, e^{\omega_1}\sin\omega_2}{1 + e^{2\omega_1} + 2\, e^{\omega_1}\cos\omega_2} - 1\right], \qquad u_2 = \frac{\sqrt{C_2}}{2\sqrt{C_1}}\left[\frac{1 - 2^{2\omega_1}}{1 + e^{2\omega_1} + 2\, e^{\omega_1}\cos\omega_2}\right]. \tag{3.42}$$

From now on, an estimate of the ω_y-component can be achieved by means of (3.15), and the evaluation of the remaining two resultants ω_x and ω_z by means of (2.1).

Let us now investigate the differential equation (2.20). We shall try to construct analytical solutions of this equation under general conditions. The conditions concern relations, wich are valid between the coefficients of (2.20), i. e. functional relations connecting the ω_z-resultant and its derivatives with the moment-components and their derivatives. We distinguish the following cases.

Case a: In this case

$$\sigma_1 = \sigma_3 , \tag{3.43}$$

or, by way of (2.21), the equation

$$(\dot\omega_z - f_3)^{\dot{}}/(\dot\omega_z - f_3) = 2\lambda_2\dot\omega_z(\dot\omega_z - f_3) , \qquad \dot\omega_z \neq f_3 , \tag{3.44}$$

holds. A simple integration of (3.43) results in

$$(\dot\omega_z - f_3)(2\lambda_2\omega_z + c_1) = -1 , \tag{3.45}$$

where c_1 is an arbitrary constant of integration. On the other hand, introducing (3.43) into (2.20) and taking into account the second of relations (2.21), after some algebra we succeed in reducing (3.20) to the linear form

$$2\dot p = 2\sigma_1 P = \sigma_1 , \qquad P = \omega_y(\dot\omega_y - f_2) .$$

Integrating the last equation and using the expression (2.4) for $\omega_y(\dot\omega_y - f_2)$, we obtain again (3.45). Consequently, we already proved that, if (3.43) is valid, then the problem under consideration is satisfied by a unique solution. Rewriting (3.45) in the Abel form

$$(2\lambda_2\omega_z + c_1)^{\dot{}}\,(2\lambda_2\omega_z + c_1) - 2\lambda_2 f_3(2\lambda_2\omega_z + c_1) = -2\lambda_2 , \tag{3.46}$$

we observe that, if the general integral of (3.46) is known, then from (2.1c) we deduce

$$\omega_x = f\omega_y , \qquad f = a_3/(J_z\dot\omega_z - M_z) ,$$

and so the two basic equations (2.1a) and (2.1b) are reduced to first-order linear differential equations, which can be analytically integrated.

Case b: Here we have conditions expressed by one of the equations

$$\sigma_1 = \sigma_2 , \qquad \sigma_1 = 0 , \qquad \sigma_2 = 0 , \qquad \sigma_3 = 0 , \tag{3.47}$$

or, by means of (2.21), one of the equations

$$\frac{(\dot\omega_z - f_3)^{\dot{}}}{\dot\omega_z - f_3} = \frac{2\dot f_2}{1 + 2f_2} , \qquad (\dot\omega_z - f_3)^{\dot{}} = 0 , \qquad \dot f_2 = f_2\frac{(\dot\omega_z - f_3)^{\dot{}}}{\dot\omega_z - f_3} , \qquad \dot\omega_z(\dot\omega_z - f_3) = 0 , \tag{3.48}$$

respectively. From all the last equations the estimation of the ω_z-component is obvious. Also, if one of equations (3.48) is valid, then the uniqueness of the solution of the problem under consideration, as well as the evaluation of the remaining two resultants ω_x and ω_y, are obtained according to the procedure developed in case a.

By now, it is interesting and useful in practice to investigate the Abel equation (3.46). For this purpose, putting

$$2\lambda_2\omega_z + c_1 = y(t), \qquad (3.49)$$

we obtain the equation

$$y\dot{y} - \alpha f_3 y = -a, \qquad (3.50)$$

where $\alpha = 2\lambda_2$ is a constant and f_3 is a known smooth function of time. Supposing, without loss of generality, that f_3 a is harmonic function, i.e. $f_3 = \overline{M}_3\, e^{i\overset{*}{\omega}t}\, l/J_z$, and taking into account that

$$\dot{y} = (\mathrm{d}y/\mathrm{d}t) = (\mathrm{d}y/\mathrm{d}f_3)\,(\mathrm{d}f_3/\mathrm{d}t) = \overset{*}{y} i\overset{*}{\omega} f_3 \qquad (3.51)$$

where the star indicates differentiation with respect to f_3, we deduce

$$y\overset{*}{y} - by = -b/f_3, \qquad b = a/i\overset{*}{\omega}. \qquad (3.52)$$

Introducing the functional transformation $y(f_3) = n(\xi)$, $\xi = bf_3$, and using the chain rule

$$\overset{*}{y} = (\mathrm{d}y/\mathrm{d}f_3) = (\mathrm{d}n/\mathrm{d}\xi)\,(\mathrm{d}\xi/\mathrm{d}f_3) = bn' \qquad (3.53)$$

in which prime means differentiation with respect to ξ, we succeed in reducing (3.52) to

$$nn' - n = -b/\xi. \qquad (3.54)$$

In the sequel by the substitution

$$n(\xi) = 1/z(\xi), \qquad (3.55)$$

equation (3.54) becomes

$$z' + z^2 - \frac{b}{\xi} z^3 = 0. \qquad (3.56)$$

The new functional transformation

$$z(\xi) = 1/s\,(\mathrm{d}\xi/\mathrm{d}s) = 1/s\dot{\xi}, \qquad (3.57)$$

where the dot respresents differentiation with respect to s, allows for bringing (3.54) into the form

$$s^2\xi\ddot{\xi} = b. \qquad (3.58)$$

Furthermore, the new substitution

$$\xi(s) = p(r), \qquad r = \ln s, \qquad (3.59)$$

by way of

$$\dot{\xi} = \mathrm{d}\xi/\mathrm{d}s = (\mathrm{d}p/\mathrm{d}r)\,(\mathrm{d}r/\mathrm{d}s) = p'/s, \qquad \ddot{\xi} = (p''/s^2) - (p'/s^2), \qquad (3.60)$$

allows for reducing (3.58) to the new form

$$p'' - p' = b/p, \qquad p \neq 0, \qquad (3.61)$$

where prime denotes differentiation with respect to r. Finally, putting

$$\begin{aligned} \overset{*}{r} &= \mathrm{d}r/\mathrm{d}q = 1/(\mathrm{d}q/\mathrm{d}r) = 1/q' = p, \\ p' &= \mathrm{d}\overset{*}{r}/\mathrm{d}r = (\mathrm{d}\overset{*}{r}/\mathrm{d}q)\,(\mathrm{d}q/\mathrm{d}r) = \overset{**}{r}/\overset{*}{r}, \qquad p'' = (\overset{**}{r}/\overset{*}{r})^*/\overset{*}{r}; \end{aligned} \qquad (3.62)$$

we transform (3.61) as follows:

$$(\overset{**}{r}/\overset{*}{r})^* - \overset{**}{r} = b. \qquad (3.63)$$

Integrating the last equation we obtain

$$\frac{\overset{**}{r}}{\overset{*}{r}} - \overset{*}{r} = bq - c_1, \qquad (3.64)$$

in which c_1 is a constant of integration. Rewriting (3.64) as

$$(\ln \overset{*}{r})^* - \overset{*}{r} = bq - c_1$$

and integrating again, we obtain

$$\ln \overset{*}{r} - r = (bq^2/2) - c_1q + \ln c_2, \qquad c_2 > 0$$

or

$$\overset{*}{r} = c_2 \exp(r) \exp[(bq^2/2) - c_1q], \tag{3.65}$$

where c_2 is a new constant of integration. This equation can be integrated by parts since it can be written as

$$\frac{dr}{\exp(r)} = c_2 \exp[(bq^2/2) - c_1q] \, dq .$$

To estimate of the right-hand side integral is not elementary. Observing by an inverse approach that the function r respresents twice the integral of the ω_z-resultant with respect to time, while q denotes twice the integral of the $(1/\omega_z)$-function with respect to time, one may write for a time-interval [0, t] the probability integral

$$e^{-r} = c_2 \int_q^\infty e^{-\frac{ia}{2\overset{*}{\omega}}q^2 - c_1q} \, dq + c_3, \tag{3.66}$$

in which b is replaced by its equivalent expression $(a/i\overset{*}{\omega})$. The estimate of the integral (3.66) is included in [15, p. 307, 3. 322.1], i.e.

$$e^{-r} = \sqrt{\frac{\pi\overset{*}{\omega}}{2a}}\, i \, e^{\sqrt{(\pi\overset{*}{\omega}i/2a)}\, c_1^2} \left[1 - \Phi\left(c_1 \sqrt{\frac{\pi\overset{*}{\omega}}{2a}}\, i + q\Big/2\sqrt{\frac{\pi\overset{*}{\omega}}{2a}\, i}\right)\right]$$

(Φ denotes the probability function), which, by means of the same reference [15, p. 1063, 9.236], can be connected with the degenerate hypergeometric function $\bar{\Phi}(a, b; z)$ as follows:

$$e^{-r} = \sqrt{\frac{\pi\overset{*}{\omega}}{2a}}\, i \, e^{\sqrt{(\pi\overset{*}{\omega}i/2a)}\, c_1^2} \frac{2\left(c_1 \sqrt{\frac{\pi\overset{*}{\omega}i}{2a}} + q\Big/2\sqrt{\frac{\pi\overset{*}{\omega}i}{2a}}\right)}{\sqrt{\pi}} \bar{\Phi}\left[\frac{1}{2}, \frac{3}{2}; c_1 \sqrt{\frac{\pi\overset{*}{\omega}i}{2a}} + q\Big/2\sqrt{\frac{\pi\overset{*}{\omega}i}{2a}}\right)^2\right]. \tag{3.67}$$

If some terms of the corresponding series representation of the degenerate hypergeometric function $\bar{\Phi}$ are selected, the evaluation of the ω_z-component can be obtained by the inverse approach.

4. Application to the case of a one-axis symmetrical gyro

A nontrivial application of practical interest in spacecraft engineering is the motion of a one-axis symmetrical gyro, subjected to a generic external loading (see [1] to [4]). For this case we shall try to construct analytical solutions of the initial dynamic system (2.1).

Assume that the gyro has symmetry such that

$$J_x = J_y < J_z \Rightarrow \alpha_3 = 0, \tag{4.1}$$

and the moment-components are arbitrary smooth functions of time. By a direct integration of equation (2.1c) we obtain

$$\omega_3(t) = F_3(t), \qquad F_3(t) = \int f_3(t)\, dt + c_1, \tag{4.2}$$

in which c_1 is a constant of integration. Consequently, equations (2.1a) and (2.1b) can be reduced to

$$\dot{\omega}_x - A_1F_3\omega_y = f_1, \qquad \dot{\omega}_y - A_2F_3\omega_x = f_2. \tag{4.3 a, b}$$

If we solve (4.3b) for ω_x, differentiate the resulting new relation once with respect to t and combine it with (4.3a), then we are led to a second-order linear differential equation with variable coefficients of the form

$$F_3\ddot{\omega}_y - \dot{F}_3\dot{\omega}_y - A_1A_2F_3^3\omega_y = Z(t), \tag{4.4}$$

in which

$$Z(t) = F_3\dot{f}_2 - \dot{F}_3f_2 + A_2F_3^3f_1. \tag{4.5}$$

Taking the homogeneous differential equation corresponding to (4.5)

$$\ddot{\omega}_{y_0} - (\dot{F}_3/F_3)\, \dot{\omega}_{y_0} - A_1A_2F_3^2\omega_{y_0} = 0 \tag{4.6}$$

and using a convenient functional transformation,

$$\omega_{y_0}(t) = \eta(\xi), \qquad \xi = \int F_3(t)\, dt, \tag{4.7}$$

we estimate

$$\dot{\omega}_{y_0} = \eta' F_3 , \qquad \ddot{\omega}_{y_0} = \eta'' F_3^2 + \eta' F_3 ,$$

where primes mean differentiations with respect to ξ. So equation (4.6) is reduced to the following equivalent (with constant coefficients) form:

$$\eta'' - A_1 A_2 \eta = 0 . \tag{4.8}$$

Since now the relations

$$A_1 = (J_y - J_z)/J_x > 0 , \qquad A_2 = (J_z - J_x)/J_y > 0$$

are valid, we estimate the general integral of (4.8) as follows:

$$\eta(\xi) = c_2 \cos (|A_1 A_2|^{1/2} \xi) + c_3 \sin (|A_1 A_2|^{1/2} \xi) ,$$

where c_2, c_3 are constants of integration. Accordingly, the general solution of (4.6) results in

$$\omega_{y_0}(t) = c_2 \cos [|A_1 A_2|^{1/2} H(t)] + c_3 \sin [|A_1 A_2|^{1/2} H(t)] \tag{4.10}$$

with

$$H(t) = \int F_3(t) \, \mathrm{d}t . \tag{4.11}$$

Finally, the general solution of the complete differential equation (4.4) is defined on the basis of the Lagrange method, namely

$$\omega_y(t) = \omega_{y_0}(t) + J_1 \cos [|A_1 A_2|^{1/2} H(t)] + J_2 \sin [|A_1 A_2|^{1/2} H(t)] , \tag{4.12}$$

where J_1, J_2 are the functionals

$$\begin{aligned} J_1(t) &= -\int Z(t) \sin [|A_1 A_2|^{1/2} H(t)] \, \mathrm{d}t / |A_1 A_2|^{1/2} H(t) , \\ J_2(t) &= \int Z(t) \cos [|A_1 A_2|^{1/2} H(t)] \, \mathrm{d}t / |A_1 A_2|^{1/2} H(t) . \end{aligned} \tag{4.13}$$

5. Conclusions

In this investigation the general decoupling and the analytical solutions, under general conditions, of the three first-order strongly nonlinear dynamic differential equations, governing the motion of a gyro, are presented by applying several convenient mathematical treatments. Among the most important results one may list the following:

i) The general decoupling method of the three beforementioned equations by way of the assumption that the moment-components are known smooth functions of time;

ii) the use of several successive functional transformations, by which analytical solutions of the decoupled differential equations are obtained, by means of general conditions in accordance with the physical problem, and

iii) the analytical solutions for the case of motion of a one-axis symmetrical gyro subjected to a general loading.

The proposed analysis is convenient and necessary for analytically solving the Euler dynamic equations. The techniques developed and the functional transformations introduced, which are original, exempt this general and complicated problem from the necessity to take recourse to numerical solutions necessitating large storage memory and complicated computer coding.

References

1 MAGNUS, K.: Kreisel-Theorie und Anwendungen (Kap. 3.4). Springer Verlag, Berlin-Heidelberg-New York 1971.

2 WITTENBURG, J.: Dynamics of systems of rigid bodies (Chap. 4.3). Teubner-Verlag, Stuttgart 1971.

3 GRAMMEL, R.: Der Kreisel, seine Theorie und seine Anwendungen. Springer-Verlag, Berlin 1950.

4 GRAMMEL, R.: Ingenieur-Arch. **22** (1954), 73–97.

5 FLÜGGE, N.: Handbook in engineering mechanics (Chap. 25). McGraw-Hill Book Co., New York 1962.

6 TIMOSHENKO, S.; YOUNG, D. H.: Advanced dynamics. McGraw-Hill Book Co., New York 1948.

7 ISHLINSKII, A.: Orientation, gyroscopes and inertial navigation (in Russian). Nauka Press, Moscow 1976.

8 IL'CHANICOV, V. P.; TERESHIN, V. G.: Integration of the equations of motion of a one-axis gyroscope on a rotating base. Mech. of Sol. (Mekh. Tver. Tela) **15** (1980) No. 1.

9 SMOL'NIKOV, B. A.: Motion of a solid acted upon by an orthogonal moment. Mech. of Sol. (Mekh. Tver. Tela) **14** (1979) No. 3.

10 FOX, G. H.; BURDERS, J. S.: The dynamically characteristics of a gyroscope with a tuned elastic suspension. J. Appl. Mech. ASME **47** (1980) 1.

11 Smol'nikov, B. A.: Generalization of Euler case of motion of a solid. PMM **51** (1967) No. 3.
12 Ormandy, D.; Maunder, L.: Dynamics of oscillogyro. J. Appl. Mech, ASME **15** (1973) No. 3.
13 Panayotounakos, D. E.; Theocaris, P. S.: On the decoupling and the solutions of the Euler kinematics equations governing the motion of a gyro. Internat. J. Non-Lin. Mech. (in print).
14 Kamke, E.: Differentialgleichungen, Lösungsmethoden und Lösungen. Bd. I. Chelsea Publ. Co., New York 1971.
15 Gradshteyn, I. S.; Ryzhik, I. M.; Table of integrals, series and products. Fourth Ed., Academic Press, New York 1965.

Received October 12, 1988

Address: Prof. Dr. D. S. Panayotounakos, Prof. Dr. Dr. P. S. Theocaris, The National Technical University of Athens, Department of Engineering Science, P. O. Box 77230, GR-175-10 Athens, Greece

ZAMM · Z. angew. Math. Mech. 70 (1990) 11, 501–507 Akademie-Verlag Berlin

GAMER, U.; MÜFTÜ, S.

On the Elastic-plastic Shrink Fit with Supercritical Interference

Dedicated to Prof. Dr.-Ing. H. LIPPMANN on the occasion of the 60th anniversary of his birthday

Die Behandlung von Preßverbänden mit idealplastischer Nabe mit Hilfe der Trescaschen Fließbedingung ist nicht möglich, wenn sowohl das Verhältnis von Außen- und Innendurchmesser als auch das Übermaß bestimmte Werte überschreiten. Weist das Nabenmaterial lineare Verfestigung auf, dann existiert eine Lösung für alle Radienverhältnisse. Es wird gezeigt, daß für kleine Verfestigungsparameter das Material in der Umgebung der Fuge einen zweifachen Wechsel der Fließbedingung durchmacht.

Shrink fits with perfectly elastic-plastic hub material cannot be treated with the help of Tresca's yield criterion, if the ratio of outer and inner diameter as well as the interference exceed certain limits. For linear hardening material, however, a solution exists for all radii ratios. It is shown that, for small hardening, the material in the neighbourhood of the interface undergoes a twofold change of the yield condition.

Нельзя обсуждать с помощью условия пластичности Треска нажимные перевязки с вполне вязко-пластическим материалом ступицы колеса, если и отношение между внешным и внутренным диаметрами и возмущение переходят несколько границ. Однако, для материала с линейным жестчением существует решение для всех отношений диаметров. Показывается, что для малово жестчения материал в окрестности стыка подвеграется двойной перемены условия пластичности.

1. Introduction

The stress distribution in elastic-plastic shrink fits, especially the most interesting pressure occurring at the interface between hub and shaft, can be determined analytically with the help of Tresca's yield criterion and the flow rule associated to it [1, 2]. In the course of assemblage, simulated simply by increasing interference, yielding starts at the bore of the hub, and a plastic region spreads into the hub. For perfectly plastic material the yield condition has the form $\sigma_\theta - \sigma_r = \sigma_0$ where σ_0 is the yield limit. However, the solution based on this form of the yield condition does not hold for arbitrary sizes of the interference. There exists, depending on the ratio of the radii of the hub, $Q := a/b$, a well defined interference for which the validity of the solution ceases. For $Q > 1/e$ and $I = 1/Q^2$, where $I := Ei/(\sigma_0 a)$ denotes nondimensional interference, the elastic region of the hub vanishes. Further growth of the interference, though not causing unresticted plastic flow, is not beneficial since it does not increase the interface pressure.

In the case $Q < 1/e$, a critical interference I_c given by

$$Q^2 I_c - \log I_c + 1 = 0 \tag{1.1}$$

exists for which the circumferential stress at the bore of the hub vanishes. An interference $I > I_c$ inserted into the solution hitherto valid predicts a negative circumferential stress at and near the inner edge of the hub thus violating the yield condition on which the solution is based. For perfectly plastic behavior, a solution of the super-critical interference problem does not exist [3]. For material with linear hardening the critical interference is determined by

$$[H + (1 + H) Q^2] I_c - \log I_c + 1 = 0 \tag{1.2}$$

where $H := \sigma_0 \eta / E$ is the scaled hardening parameter [4]. This equation has two real roots if the content of the bracket is smaller than $1/e^2$ or

$$H < \frac{(1/e^2) - Q^2}{1 + Q^2}. \tag{1.3}$$

Of these, the smaller one is physically relevant. For larger values of H, due to hardening, the circumferential stress does not vanish in the hub for any interference.

The shrink fit with supercritical interference is treated in [3]. With the interference surpassing the critical one given by (1.2), the second stage of plastic flow starts. The yield condition governing the plastic flow of the hub material at the interface changes into $\sigma_r = -\sigma_Y$, and a second plastic zone spreads, in which the circumferential stress vanishes and plastic flow occurs in all three principal directions. But again, this solution does not hold for arbitrarily large interferences. The end of the second stage of plastic flow is characterized by a stationary value of the axial plastic strain at the bore of the hub. Larger values of the interferences inserted into the equations governing the second stage of plastic flow predict a decrease in axial plastic strain at and near the inner edge of the hub. But plastic strain cannot decrease unless flow in the opposite direction occurs. Thus, the third stage of plastic flow must begin.

As in the related case of the pressure loaded hole, the yield condition at the interface switches back to its previous form valid throughout during the first stage of plastic flow and later in the outer zone of the plastic region [5]. As a consequence of this, the plastic strain in axial direction does not change any more. One finds that during the third stage of plastic flow the fast growing third plastic zone absorbs the slower expanding second plastic zone. The latter vanishes completely at the end of the third stage. The following fourth stage of plastic flow differs from the first stage only in the fact that in a certain part of the hub a plastic predeformation in axial direction exists. With still increasing interference, the hub finally becomes fully plasticized, and the fifth stage of plastic flow starts. Fig. 1 shows schematically the border radii as a function of the interference. As in other problems of elastic-plastic stress analysis based on Tresca's yield criterion, the assumptions leading to these predictions must be verified by numerical evidence.

This paper is devoted to the study of the third, fourth and fifth stage of plastic flow not dealt with previously. As usual in the treatment of shrink fits, it is assumed that the solid shaft is in a state of plane stress and remains elastic throughout the assemblage process.

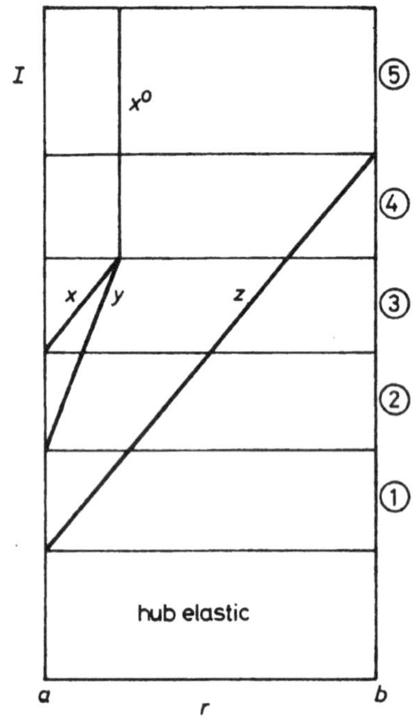

Fig. 1. Sketch of border radii as a function of interference

2. The second stage of the plastic flow

During the second stage of plastic flow the hub consists of the elastic region near the outer edge and the plastic region. The latter is composed of the outer plastic zone and of the inner plastic zone adjacent to the interface of the shrink fit. The general expressions for stresses, displacement and plastic strains in shaft and hub read [3]:

shaft, $0 \leqq r \leqq a$,

$$\bar{\sigma}_r = \bar{\sigma}_\theta = -\bar{p}\,, \tag{2.1}$$

with $\bar{\sigma}_{ij} := \sigma_{ij}/\sigma_0$,

$$\frac{E}{\sigma_0} u = -(1-\nu)\,\bar{p}r\,, \tag{2.2}$$

elastic region, $z \leqq r \leqq b$,

$$\bar{\sigma}_r = -C_1\left(\frac{1}{r^2} - \frac{1}{b^2}\right), \qquad \bar{\sigma}_\theta = C_1\left(\frac{1}{r^2} + \frac{1}{b^2}\right), \tag{2.3), (2.4}$$

$$\frac{E}{\sigma_0} u = (1-\nu)\,\bar{\sigma}_r r + \frac{2C_1}{r} = C_1\left[(1+\nu)\frac{1}{r^2} + (1-\nu)\frac{1}{b^2}\right] r\,, \tag{2.5}$$

first plastic zone, $y \leqq r \leqq z$,

$$\bar{\sigma}_r = \frac{1}{1+H}\left(\log r - \frac{H}{2}\frac{C_4}{r^2}\right) + C_3\,, \tag{2.6}$$

$$\bar{\sigma}_\theta = \frac{1}{1+H}\left(\log r + 1 + \frac{H}{2}\frac{C_4}{r^2}\right) + C_3\,, \tag{2.7}$$

$$\frac{E}{\sigma_0} u = (1 - \nu) \bar{\sigma}_r r + \frac{C_4}{r} = \frac{1 - \nu}{1 + H} r \log r + \left[1 - \frac{(1 - \nu) H}{2(1 + H)}\right] \frac{C_4}{r} + (1 - \nu) C_3 r, \tag{2.8}$$

$$\bar{\varepsilon}_\theta^p = -\bar{\varepsilon}_r^p = \bar{\sigma}_r - \bar{\sigma}_\theta + \frac{C_4}{r^2} = \frac{1}{1 + H}\left(\frac{C_4}{r^2} - 1\right) \tag{2.9}$$

with $\bar{\varepsilon}_{ij}^p = E\varepsilon_{ij}^p/\sigma_0$,

second plastic zone, $a \leqq r \leqq y$,

$$\bar{\sigma}_r = \frac{C_5}{r}, \qquad \bar{\sigma}_\theta = 0, \tag{2.10), (2.11}$$

$$\frac{E}{\sigma_0} u = \frac{1}{H} r + \frac{1 + H}{H} C_5 \log r + C_6, \tag{2.12}$$

$$\bar{\varepsilon}_r^p = \frac{1}{H}\left(1 + \frac{C_5}{r}\right) \tag{2.13}$$

$$\bar{\varepsilon}_\theta^p = \frac{1}{H} + \left(\frac{1 + H}{H} \log r + \nu\right) \frac{C_5}{r} + \frac{C_6}{r}, \tag{2.14}$$

$$\bar{\varepsilon}_z^p = -\frac{2}{H} - \left(\frac{1 + H}{H} \log r + \frac{1}{H} + \nu\right) \frac{C_5}{r} - \frac{C_6}{r}. \tag{2.15}$$

These expressions contain the unknowns p, C_1, C_3, C_4, C_5, C_6. Further unknowns are the border radii z and y. Making use of the requirement of continuity of stresses and displacements at the border radii $r = z$ and $r = y$, one obtains

$$C_1 = \frac{z^2}{2}, \qquad C_3 = -\frac{1}{2(1 + H)}\left[2 \log y + H \frac{z^2}{y^2} + 2\right], \tag{2.16), (2.17}$$

$$C_4 = z^2, \qquad C_5 = -\frac{1}{1 + H}\left(H \frac{z^2}{y^2} + 1\right) y, \tag{2.18), (2.19}$$

$$C_6 = \left\{\frac{1}{1 + H}\left[\frac{1 + H}{H} \log y + \frac{1}{H} + \nu\right]\left(H \frac{z^2}{y^2} + 1\right) - \frac{2}{H}\right\} y \tag{2.20}$$

and

$$(1 + H) \frac{z^2}{b^2} = \log \frac{z^2}{y^2} - H \frac{z^2}{y^2} - 1. \tag{2.21}$$

Note, that the equations and conditions used to determine the above results hold without modification in the third stage of plastic flow also!

To complete the results of the second stage of plastic flow one makes use of the continuity of radial stress at the interface $r = a$ and of the geometrical condition

$$u^{\mathrm{H}} - u^{\mathrm{S}} = i, \tag{2.22}$$

according to which the difference of displacement of hub and shaft at the interface equals the interference. There comes

$$\bar{p} = \frac{1}{1 + H}\left(H \frac{z^2}{y^2} + 1\right) \frac{y}{a} = -\frac{C_5}{a}, \tag{2.23}$$

$$I = \left[\frac{1}{H}\left(H \frac{z^2}{y^2} + 1\right) \log \frac{y}{a} + \frac{z^2}{y^2}\right] \frac{y}{a} - \frac{1}{H}\left(\frac{y}{a} - 1\right). \tag{2.24}$$

For the numerical evaluation, one uses $L := z^2/y^2$ as a substitute load parameter. Choosing a value L from the interval $L_2 \leqq L \leqq L_3$, one obtains with the help of (2.21) z and y and therewith all the other unknowns including the actual load parameter I. L_2 equals the critical interference I_c, and L_3 will be determined now.

The axial plastic strain in the second plastic region $a \leqq r \leqq y$ reads

$$\bar{\varepsilon}_z^p = \frac{1}{H}(1 + HL) \frac{Y}{R} \log \frac{R}{Y} + \frac{2}{H}\left(\frac{Y}{R} - 1\right) \tag{2.25}$$

with $Y := y/a$ and $R := r/a$. Disregarding its range of validity, one finds that, for a fixed R, $\bar{\varepsilon}_z^p$ grows with L to a maximum and drops thereafter. This happens first at the interface $r = a$. Hence the second stage of plastic flow ends for L_3 given by $\mathrm{d}\bar{\varepsilon}_z^p/\mathrm{d}L = 0$ at $R = 1$ with

$$\frac{\mathrm{d}\bar{\varepsilon}_z^p}{\mathrm{d}L} = \frac{Y}{R} \log \frac{R}{Y} + \frac{1}{HR} \left[(1 + HL) \left(\log \frac{R}{Y} - 1 \right) + 2 \right] \frac{\mathrm{d}Y}{\mathrm{d}L}, \tag{2.26}$$

$$Y = \left[\frac{\log L - HL - 1}{(1 + H) Q^2 L} \right]^{1/2}, \qquad \frac{\mathrm{d}Y}{\mathrm{d}L} = \frac{2 - \log L}{2(1 + H) Q^2 Y L^2}. \tag{2.27), (2.28}$$

Numerical results for the distribution of stresses, displacement and plastic strains are given in [3].

3. The third stage of plastic flow

It is expected that in the third stage of plastic flow the circumferential stress at and near the interface will be tensile again [5]. The numerical results verify this conjecture later on. A detailed derivation of stresses, displacement and plastic strains in the third plastic region $a \leqq r \leqq x$ is given now.

The yield condition reads

$$\sigma_\theta - \sigma_r = \sigma_Y . \tag{3.1}$$

Associated to it is the flow rule

$$\mathrm{d}\varepsilon_\theta^p / \mathrm{d}\varepsilon_r^p = -1 , \qquad \mathrm{d}\varepsilon_z^p = 0 . \tag{3.2}$$

Integration (with respect to the load parameter) results in

$$\varepsilon_z^p = \varepsilon_z^{per}(r) , \qquad \varepsilon_r^p + \varepsilon_\theta^p = -\varepsilon_z^{per} . \tag{3.3}$$

The axial plastic strain, accumulated at a material point in the second plastic zone, grows to a stationary value. At this instant, the border radius x passes the material point under consideration, and the axial plastic strain is frozen to permanent plastic strain.

The displacement is derived in the following way: The sum of the radial and circumferential strain is given by

$$\varepsilon_r + \varepsilon_\theta = \varepsilon_r^e + \varepsilon_\theta^e - \varepsilon_z^{per} . \tag{3.4}$$

Making use of the geometrical relations,

$$\varepsilon_r = \frac{\mathrm{d}u}{\mathrm{d}r}, \qquad \varepsilon_\theta = \frac{u}{r}, \tag{3.5}$$

and Hooke's law,

$$E\varepsilon_r^e = \sigma_r - \nu\sigma_\theta , \qquad E\varepsilon_\theta^e = \sigma_\theta - \nu\sigma_r , \tag{3.6}$$

one obtains

$$E \left(\frac{\mathrm{d}u}{\mathrm{d}r} + \frac{u}{r} \right) = (1 - \nu) (\sigma_r + \sigma_\theta) - E\varepsilon_z^{per} . \tag{3.7}$$

Now, the circumferential stress is eliminated with the help of the equation of equilibrium,

$$\frac{\mathrm{d}\sigma_r}{\mathrm{d}r} + \frac{\sigma_r - \sigma_\theta}{r} = 0 , \tag{3.8}$$

and integration yields

$$\frac{E}{\sigma_0} u = (1 - \nu) \bar{\sigma}_r r - \frac{1}{r} \int \bar{\varepsilon}_z^{per} r \, \mathrm{d}r + \frac{C_7}{r}. \tag{3.9}$$

Therefrom, one derives

$$\bar{\varepsilon}_\theta^p = \bar{\sigma}_r - \bar{\sigma}_\theta - \frac{1}{r^2} \int \bar{\varepsilon}_z^{per} r \, \mathrm{d}r + \frac{C_7}{r^2}, \tag{3.10}$$

$$\bar{\varepsilon}_r^p = \bar{\sigma}_\theta - \bar{\sigma}_r + \frac{1}{r^2} \int \bar{\varepsilon}_z^{per} r \, \mathrm{d}r - \bar{\varepsilon}_z^{per} - \frac{C_7}{r^2}. \tag{3.11}$$

In order to find the stresses one has first to determine the equivalent plastic strain entering the hardening law. The equivalence of the increment of plastic work,

$$\sigma_r \, d\varepsilon_r^p + \sigma_\theta \, d\varepsilon_\theta^p + \sigma_z \, d\varepsilon_z^p = \sigma_Y \, d\varepsilon_{EQ}, \tag{3.12}$$

yields

$$-d\varepsilon_r^p = d\varepsilon_{EQ}. \tag{3.13}$$

The same relation holds in the first and in the second plastic zone. This means that the increment of equivalent plastic strain at a material point near the interface, coming after onset of yield into the first plastic zone, then into the second plastic zone and finally into the third plastic zone, equals the negative increment of the radial plastic strain throughout the plastic deformation. Therefore

$$\varepsilon_{EQ} = -\varepsilon_r^p. \tag{3.14}$$

Inserting the equivalent plastic strain given by (3.11) with the yield condition (3.1) into the hardening law,

$$\bar{\sigma}_Y = 1 + H\bar{\varepsilon}_{EQ}, \tag{3.15}$$

one obtains the yield stress $\bar{\sigma}_Y$ as a function of radius. Therewith integration of the equilibrium condition (3.8) yields

$$\bar{\sigma}_r = \frac{1}{1+H} \log r + \frac{H}{2(1+H)} \left[\frac{1}{r^2} \int \bar{\varepsilon}_z^{per} r \, dr + \int \frac{\bar{\varepsilon}_z^{per}}{r} \, dr - \frac{C_7}{r^2} \right] + C_8. \tag{3.16}$$

Adding to this the yield stress $\bar{\sigma}_Y$, one obtains

$$\bar{\sigma}_\theta = \frac{1}{1+H} (\log r + 1) + \frac{H}{2(1+H)} \left[-\frac{1}{r^2} \int \bar{\varepsilon}_z^{per} r \, dr + \int \frac{\bar{\varepsilon}_z^{per}}{r} \, dr + 2\bar{\varepsilon}_z^{per} + \frac{C_7}{r^2} \right] + C_8. \tag{3.17}$$

For $\bar{\varepsilon}_z^{per} = 0$ (no plastic predeformation) all these results agree with the equations governing the first plastic region.

In these expressions the two additional unknowns C_7 and C_8 occur. Also unknown is the border radius x separating the third plastic region from the second plastic region. It has to be emphasized that the results (2.16) to (2.20) for the integration constants C_1 to C_6 remain valid without modification. The same holds for (2.21) connecting the border radii z and y. The pressure $\bar{p}$ and the interference I, however, have to be determined anew. The border radius x as a function of the load parameter L is found in the following way: From $d\bar{\varepsilon}_z^p/dL = 0$ according to (2.26), one obtains the function f assigning to a radius R the load parameter L^* for which the axial plastic strain assumes a stationary value at this radius, $L^* = f(R)$. Its inverse f^{-1} assigns to the load parameter L the border radius X, $X = f^{-1}(L)$. Inserting $L^* = f(R)$ into (2.25), one finds the permanent plastic strain

$$\bar{\varepsilon}_z^{per} = \frac{1}{H} (1 + HL^*) \frac{Y(L^*)}{R} \log \frac{R}{Y(L^*)} + \frac{2}{H} \left(\frac{Y(L^*)}{R} - 1 \right) \tag{3.18}$$

as a function of the radius R.

From the continuity of stresses at $r = x$ there follows

$$C_7 = \left[\frac{1}{H} \left(H \frac{z^2}{y^2} + 1 \right) \frac{y}{x} + \frac{1}{x^2} INT_1(x) - \bar{\varepsilon}_z^{per}(x) - \frac{1}{H} \right] x^2, \tag{3.19}$$

$$C_8 = -\frac{1}{2(1+H)} \left[\left(H \frac{z^2}{y^2} + 1 \right) \frac{y}{x} + 2 \log x + H \left(INT_2(x) + \bar{\varepsilon}_z^{per}(x) \right) + 1 \right] \tag{3.20}$$

with $INT_1() = \int_a^{()} \frac{\bar{\varepsilon}_z^{per}}{r} \, dr$ and $INT_2() = \int_a^{()} \frac{\bar{\varepsilon}_z^{per}}{r} \, dr$. Finally, the continuity of radial stress at the interface $r = a$ and the geometrical condition (2.22) yield

$$\bar{p} = \frac{1}{1+H} \log \frac{x}{a} + \frac{H}{2(1+H)} \left[\frac{1}{H} \left(H \frac{z^2}{y^2} + 1 \right) \frac{y}{x} \left(\frac{x^2}{a^2} + 1 \right) + \frac{1}{a^2} INT_1(x) \right.$$
$$\left. + INT_2(x) - \left(\bar{\varepsilon}_z^{per}(x) + \frac{1}{H} \right) \left(\frac{x^2}{a^2} - 1 \right) \right], \tag{3.21}$$

$$I = \left[\frac{1}{H} \left(H \frac{z^2}{y^2} + 1 \right) \frac{y}{x} + \frac{1}{x^2} INT_1(x) - \bar{\varepsilon}_z^{per}(x) - \frac{1}{H} \right] \frac{x^2}{a^2} = \frac{C_7}{a^2}. \tag{3.22}$$

The third stage of plastic flow comes to an end when the border radius x reaches y. Their common final value is denoted by x^0. From $d\bar{\varepsilon}_z^p/dL = 0$ at $R = Y$, using (2.26) and (2.28), one obtains $L_4 = e^2$. (The second value $L_4 = 1/H$ is physically irrelevant.)

The question arises whether the hub can be completely plasticized during the third stage of plastic flow. The answer is negative. Inserting into (2.21) the largest value of L that occurs during the third stage, $L_4 = e^2$, one obtains

$$\frac{z^2}{b^2} = \frac{1 - He^2}{1 + H} < 1 . \tag{3.23}$$

Thus, at the end of the third stage, an elastic region will still be left.

For $\nu = 0.3$, $H = 0.04$, $Q = 0.25$ and the border radius $X = 1.1$ Fig. 2 shows the stresses in the hub and Fig. 3 the displacement and plastic strains as a function of radius. The circumferential stress in the third plastic region, $1 \leqq R \leqq X$, is too small to be visible in Fig. 2. It is plotted in magnified scale on Fig. 4. To $X = 1.1$ there belong $L = Z^2/Y^2 = 6.969$ and $I = 10.331$.

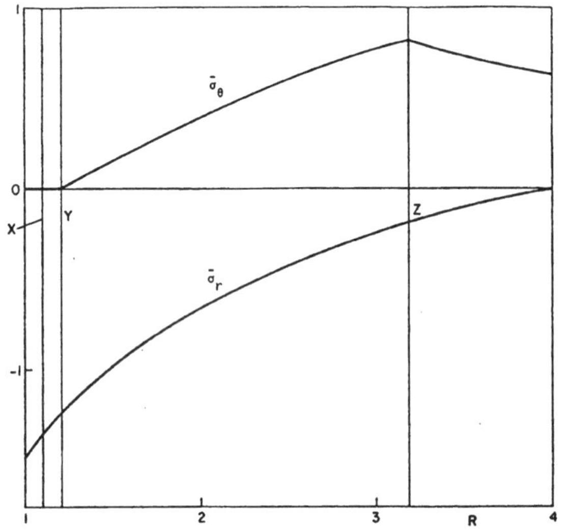

Fig. 2. Stresses in hub as functions of radius

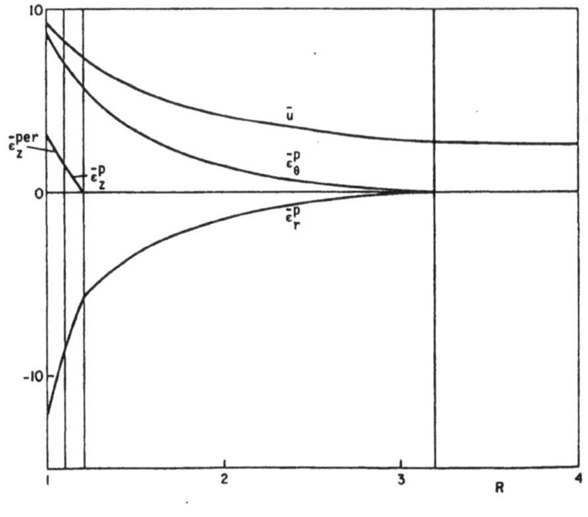

Fig. 3. Displacement and plastic strains in hub as functions of radius

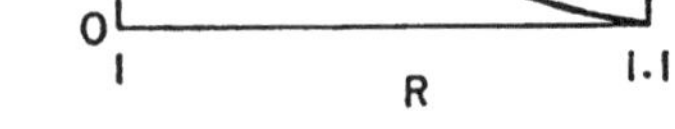

Fig. 4. Circumferential stress in third plastic region as function of radius

4. The fourth and fifth stage of plastic flow

The fourth stage of plastic flow is a generalization of the first stage. The difference consists in the fact that in the fourth stage there exists a permanent plastic strain, $\bar{\varepsilon}_z^{per}$, in the zone $a \leqq r \leqq x^0$, which was absent in the first stage. Stresses and displacement in the elastic region, $z \leqq r \leqq b$, are still given by (2.1) and (2.2). In the entire plastic region, $a \leqq r \leqq z$, the expressions derived for the third plastic region hold. From continuity requirements and the geometrical condition one obtains

$$C_1 = \frac{z^2}{2}, \qquad C_7 = z^2 + INT_1(x^0), \tag{4.1), (4.2}$$

$$C_8 = -\frac{1}{1+H}\log z - \frac{1}{2}\left(1 - \frac{z^2}{b^2}\right) + \frac{H}{2(1+H)}\left[1 - INT_2(x^0)\right], \tag{4.3}$$

$$\bar{p} = \frac{1}{1+H}\log\frac{z}{a} + \frac{1}{2}\left(1 - \frac{z^2}{b^2}\right) + \frac{H}{2(1+H)}\left[\frac{z^2}{a^2} - 1 + \frac{1}{a^2}INT_1(x^0) + INT_2(x^0)\right], \tag{4.4}$$

$$I = \frac{z^2}{a^2} + \frac{1}{a^2}INT_1(x^0). \tag{4.5}$$

For $\bar{\varepsilon}_z^{per} = 0$, these expressions agree with the ones pertinent to the partially plasticized hub of a shrink fit with subcritical interference [4]. The fourth stage of plastic flow ends with

$$I_5 = Q^2 + \frac{1}{a^2}INT_1(x^0). \tag{4.6}$$

For this interference the hub is completely plasticized.

In the final state of plastic flow the elastic region of the hub has disappeared. From the boundary condition and the condition of continuity imposed on the radial stress and from the geometrical condition there follow

$$C_7 = Ia^2, \tag{4.7}$$

$$C_8 = -\frac{1}{1+H}\log b + \frac{H}{2(1+H)}\left[IQ^2 - \frac{1}{b^2}INT_1(x^0) - INT_2(x^0)\right], \tag{4.8}$$

$$\bar{p} = -\frac{1}{1+H}\log Q + \frac{H}{2(1+H)}\left[I(1 - Q^2) + \frac{1}{b^2}INT_1(x^0) + INT_2(x^0)\right]. \tag{4.9}$$

Again, this is a generalization of the fully plasticized hub of a shrink fit with subcritical interference [4]. It should be borne in mind that these results, being based on the assumption of small strain, do not hold for arbitrarily large interferences.

References

1 KOLLMANN, F. G.: Die Auslegung elastisch-plastisch beanspruchter Querpreßverbände. Forsch. Ing.-Wes. **44** (1978), 1–11.
2 KOLLMANN, F. G.: Welle-Nabe-Verbindungen. Konstruktionsbücher Bd. 32. Springer, Berlin 1984.
3 GAMER, U.: The elastic-plastic shrink fit with supercritical interference. Acta Mech. **61** (1986), 1–14.
4 GAMER, U.; LANCE, R. H.: Elastisch-plastische Spannungen im Schrumpfsitz. Forsch. Ing.-Wes. **48** (1982), 192–198.
5 GAMER, U.: Die elastisch-plastische Scheibe mit druckbelasteter Bohrung. Forsch. Ing.-Wes. **54** (1988), 94–98.

Received February 10, 1989

Addresses: Prof. Dr. UDO GAMER, Institut für Mechanik, Technische Universität Wien, Wiedner Hauptstraße 8–10, A-1040 Wien, Österreich; SINAN MÜFTÜ B. Eng., Graduate Student, Department of Mechanical Engineering, Middle East Technical University, 06531 Ankara, Turkey

BOOK REVIEWS

Moss, F.; McClintock, P. V. E. (eds.), Noise in Nonlinear Dynamical Systems. Vol. 1: Theory of Continuous Fokker-Planck Systems. Vol. 2: Theory of Noise Induced Processes in Special Applications. Vol. 3: Experiments and Stimulations. Cambridge etc., Cambridge University Press 1989. XVI, 353 pp.; XVIII, 388 pp.; XVI, 278 pp.; £ 50.00 H/b; £ 50.00 H/b; £ 45.00 H/b. 3 Vol. set £ 120.00 H/b. ISBN 0-521-35228-2; 0-521-35229-0; 0-521-35265-7; 3 Vol. set 0-521-37117-1

Zu den in Titel und Untertiteln umrissenen Themen liefern eine Reihe von Autoren etwa dreißig selbständige Beiträge über die wichtigsten Ergebnisse ihres eigenen Arbeitsgebiets. Nach diesem Konzept ist es den Herausgebern gelungen, zu den fundamentalen Problemen dynamischer Systeme mit unterschiedlichen Typen von Nichtlinearitäten sowie von Rausch- und Dissipationseigenschaften aktuelle Übersichtsarbeiten mit wesentlichen Originalanteilen vorzulegen.

Band 1 ist charakterisierbar nach den verwendeten Methoden (Markov-Theorie, Projektions- und Funktionalmethoden) und den behandelten Erscheinungen (Stochastik, Bistabilität und Bifurkationen, Übergangs- und Evolutionsphänomene). Im Band 2 werden die Effekte diskreter Dynamiken, Quanteneffekte, Phasenübergänge und Symmetriebrechungen untersucht, bereits mit ersten Anwendungen auf optische Systeme. Band 3 schließt die Thematik nach der experimentellen und numerischen Seite hin ab, und zwar in einigen konkreten Problemen: Turbulenz in He II, Wachstumsprozesse in Flüssigkristallen, Irregularitäten in Farbstofflasern, Simulationen zu speziellen Problemen stochastisch beeinflußter nichtlinearer Dynamik.

Ausführliche Literaturangaben zu jedem Beitrag und ein instruktives Stichwortverzeichnis zu jedem Band erleichtern das Auffinden von mathematischen und physikalischen Querverbindungen in der großen Vielfalt der Einzelerscheinungen und Behandlungsaspekte.

Insgesamt kann die vorliegende Reihe als ein modernes und autorisiertes Kompendium über die aufregend vielfältigen Formen des Zusammenwirkens stochastischer und nichtlinearer Irregularitäten betrachtet werden, auch im Hinblick auf Fragen und Grenzen der Vorhersagbarkeit bei realen dynamischen Systemen von steigender praktischer und technischer Bedeutung.

Auch der historischen Entwicklung des Gebiets ist Rechnung getragen worden. So enthält Band 1 die erste englische Übersetzung der Pionierarbeit von Pontriagin, Andronov und Vitt von 1933. Schwer zu realisieren, aber von äußerstem Interesse wäre ein Konsens der in dieser Reihe vereinigten hervorragenden Vertreter ihres Gebiets über die vermutlich weitere Entwicklung dieser Untersuchungen makroskopischer Systeme, insgesamt oder nach verschiedenen Richtungen.

Berlin G. Wallis

Chulaevsky, V. A., Almost Periodic Operators and Related Nonlinear Integrable Systems. Manchester/New York, Manchester University Press 1989. X, 105 pp., £ 29.95. ISBN 0-7190-2667-9 (Nonlinear Science: Theory and Applications)

Die Hillsche Gleichung ist Ausgangspunkt zahlreicher mathematischer Untersuchungen und verschiedener physikalischer Anwendungen. So ist z. B. die Mathieusche Gleichung ein Spezialfall hiervon, und die parametrisch erregte Resonanz mit ihren „Zungen" sowie das quantentheoretische Verhalten eines Teilchens im periodischen Potential sind wohl die bekanntesten Beispiele hierzu.

Grundlegend für derartige Betrachtungen ist die Theorie periodischer Schrödingeroperatoren, die zum Blochschen Theorem führt, bei dem die Lösungen der Schrödingergleichung durch Blochfunktionen dargestellt sind und der darin auftretende Quasiimpuls innerhalb der „erlaubten" Energiebänder reell, außerhalb jedoch komplex ist. Eine wichtige Rolle spielt dabei die integrale Zustandsdichte, und die spektrale Zerlegung derartiger Operatoren führt auf ein absolut stetiges Spektrum.

Jedoch ist es erforderlich, die Betrachtungen auf diejenigen Fälle auszudehnen, wo die Periodizität gestört ist, die physikalisch von Kristallen mit Fehlstellen bis zu Quasikristallen oder amorphen Substanzen reichen.

Derartige Überlegungen führen zum Begriff der fastperiodischen Schrödingeroperatoren. Die integrale Zustandsdichte ist weiterhin eine brauchbare Größe, und mit ihrer Hilfe läßt sich das Phänomen der „Selbstmittelung" erfassen, eine Art ergodisches Theorem. Die „grenzperiodischen" Funktionen, Potentiale und Operatoren verkörpern den ersten Schritt einer systematischen Erweiterung des periodischen Falles. Es ist möglich, zu zeigen, daß hier ebenfalls ein absolut stetiges Spektrum vorliegt, wenn nur eine schwache Störung auftritt.

Die Abhandlungen über „quasiperiodische" Schrödingeroperatoren sind wesentlicher Bestandteil der Darlegung. In diesem Fall ist eine komplizierte Gestalt des Spektrums vorhanden; in einigen Parameterbereichen finden wir das Blochsche Verhalten, in anderen begegnen wir der bekannten Anderson-Lokalisierung, die typisch für zufällige Potentialverteilungen ist.

Vielversprechend ist der Ausblick auf Zusammenhänge zwischen der Spektralanalysis periodischer Operatoren und nichtlinearen integrablen Systemen.

Die hier vorliegenden Ausführungen geben einen instruktiven Einblick in den gegenwärtigen Stand dieser aktuellen Thematik.

Dresden P. Möbius

De Swart, H. E., Vacillation and Predictability Properties of Low-Order Atmospheric Spectral Modes. Amsterdam, Stichting Mathematisch Centrum 1989. 121 pp., Dfl. 19.00 ISBN 90-6196-368-0(CWI Tract 60)

Die Gleichungen der atmosphärischen Zirkulation und der quasigeostrophischen Bewegung, zurückgehend auf die „multiple-scale-analysis" von Pedlosky (1984), werden in Modellen mit 3, 6 und 10 Moden sowie unter Einbeziehung weißen, farbigen und deterministischen Rauschens untersucht. Bifurkationen periodischer Lösungen zu chaotischen über homokline Orbits werden systematisch behandelt und erlauben ein Analyse barokliner Instabilitäten als einen wichtigen nichtlinearen Mechanismus der Atmosphäre.

Durch stochastische Einflüsse erzwungene oder getriggerte Effekte werden im Hinblick auf Klassifikation und Charakterisierung von Wetterregimen behandelt. Meteorologische Vorhersageverfahren durch Einbeziehung von Methoden der Lyapunov-Exponenten und von Seltsamen Attraktoren, d. h. durch die Methoden der modernen Chaos-Theorie zu verbessern, ist Inhalt des letzten Kapitels und erklärtes Ziel des vorliegenden „CWI-Tracts" des „Centrums voor Wiskunde en Informatica".

Zu einem mathematisch interessanten Problem, bei dem durch die Untersuchungen von Lorenz die moderne Chaostherie ihren Ausgang nahm und das inzwischen auch eine große praktische Bedeutung erlangt hat, werden von de Swart wichtige Ergebnisse aus der „Netherlands Foundation of the Technical Sciences" in einer wohldurchdachten und übersichtlichen Form vorgelegt. Bei der Komplexität der analysierten Zusammenhänge ist diese mathematisch fundierte Systematik von besonderer Bedeutung. So gehen die möglichen Anwendungen über die Wettervorhersage weit hinaus und erstrecken sich auch auf Teilgebiete der Turbulenzforschung sowie der Spektral- und Zeitreihen-Analyse nichtlinearer dynamischer Systeme.

Berlin G. Wallis

ZAMM · Z. angew. Math. Mech. 70 (1990) 11, 509–516 Akademie-Verlag Berlin

MERIC, R. A.

Shape Variations for Viscoelastic Solids Under Transient Response

Sensitivitätsausdrücke für ein allgemeines Gütekriterium (z. B. ein Integralfunktional) hinsichtlich der Form sind für einen viskoelastischen Festkörper im Übergangsverhalten abgeleitet worden. Differential- und Integraltypen der Materialgleichungen werden in der Arbeit betrachtet, wobei eine Kontinuumsnäherung angenommen wird. Unter Verwendung der substantiellen Ableitung und der Methode der adjungierten Variablen wird die Sensitivität des allgemeinen Integralfunktionals in Termen der ursprünglichen und adjungierten Funktionen erhalten. Es wird auch ein einfaches Beispiel gegeben, um die Resultate durch Vergleich mit jenen, die durch direkte Differentiation gefunden wurden, zu prüfen.

Sensitivity expressions for a general performance criterion (cf. an integral functional) with respect to shape have been derived for viscoelastic solids under transient response. Differential and integral types of constitutive equations are considered in the study, in which a continuum approach is adopted. By using the material derivative concept and the adjoint variable method the sensitivity of the general integral functional is obtained in terms of primary and adjoint functions. A simple example problem is also provided to check the results by comparing them with those found by a direct differentiation.

Выводятся выражения чувствительности для общего критерия качества (на пример для интегрального функционала) относительно формы для вязкоупругого твердего тела при переходной характеристике. В статье рассматриваются дифференциальные и интегральные рода уравнений консистенции. Здесь принимается континуум-подход. Применив концепцию субстанциальной производной и метод сопряжиенных переменных получается чувствительность общего интегрального функционала на языке первичных и сопряженных функций. Кроме того обеспечивается простая примерная задача чтобы проверять результаты путем сравнении с результатами полученными при прямом дифференцировании.

1. Introduction

Sensitivity analysis (SA) of integral functionals, representing a system's response, with respect to continuous functions (i.e., the so-called decision variables) plays a very important role in the solution of inverse problems, such as identification, optimization, optimal control problems, etc. Such an analysis usually provides gradient information for first-order nonlinear programming methods of solution. In particular, optimal designs of structures require SA performed with respect to shape, i.e., the decision variable is the shape configuration itself. Shape SA results have been obtained successfully for elastic structures in the context of structural optimization [1–3]. The material derivative concept and the adjoint variable method are usually used in the continuum approach in order to find the sensitivity of functionals, representing either objective functions or behavioural constraints. Such procedures have recently been extended for the shape SA of elastic structures under dynamic loads in [4].

Optimal design of viscoelastic solids, on the other hand, was considered by LEKSZYCKI and OLHOFF [5] and PIERSON [6] where the optimal distributions of viscoelastic damping layers on elastic beams were studied. In [5–6], one-dimensional beams performing transverse vibrations under given harmonically varying external loads were investigated. In other words, for the one-dimensional composite beams only the steady-state vibrations were taken into account. As for the constitutive equations of the viscoelastic solids differential forms were adopted.

The present study addresses the question of shape SA of viscoelastic solids incorporating the following features: i) three-dimensional solids are considered, ii) full dynamic (including transient) effects are taken into account, and iii) both differential and integral representations are adopted for the material behavior. In the following sections, a three-parameter differential form of the constitutive equations will first be investigated, followed by a general integral form incorporating the memory effects. A simple example will also be provided to check the shape SA results analytically.

2. Standard linear model of differential type

2.1. Primary problem

In this section, the linear, isotropic, homogeneous viscoelastic solid will be taken as obeying the three-parameter constitutive equations of differential type (cf., standard linear model) as follows:

$$\sigma_{ij} + \tau\dot{\sigma}_{ij} = k_0 t_{ij} + k_v \dot{t}_{ij} \qquad (1)$$

where σ_{ij} is the stress tensor; the superpositioned dot refers to time differentiation; τ, k_0 and k_v are material properties; and t_{ij} is given by

$$t_{ij} = \lambda\theta\delta_{ij} + 2\mu\varepsilon_{ij} \qquad (2)$$

where λ and μ are Lamé's constants; δ_{ij} is the Kronecker delta; ε_{ij} is the strain tensor, which is related to the displacements u_i through the kinematic conditions as

$$\varepsilon_{ij} = \tfrac{1}{2}(u_{i,j} + u_{j,i}) \tag{3}$$

where $(.)_{,i}$ refers to the differention of (.) with respect to x_i; and θ is the trace of the strain tensor given by

$$\theta = \varepsilon_{11} + \varepsilon_{22} + \varepsilon_{33} .$$

By appropriately taking limiting values for the material properties in equation (1) the standard linear viscoelastic solid may be reduced to a Kelvin-Voigt solid (with $\tau = 0$ and $k_0 = 1$) and a Maxwell solid (with $k_0 = 0$ and $k_v = 1$) [7]. It will be convenient to work with the tensor p_{ij} defined as

$$p_{ij} = \sigma_{ij} + \tau\dot{\sigma}_{ij}$$

which reduces to the stress tensor σ_{ij} when $\tau = 0$. It is noted that the memory dependence of the media is characterized by a differential (instead of an integral) equation in time, i.e., equation (1).

Equations of motion in terms of the stress tensor σ_{ij}, body force vector b_i, displacements u_i, and density ϱ may be written as

$$\text{in } V\colon \quad \sigma_{ij,j} + b_i = \varrho\ddot{u}_i . \tag{4}$$

Mixed boundary conditions may be imposed such that

$$\text{on } S_u\colon \quad u_i = u_i^0 , \tag{5}$$

$$\text{on } S_p\colon \quad p_i = p_{ij}n_j = p_i^0 \tag{6}$$

where S_u and S_p are parts of the bounding surface S of V (i.e., $S = S_u \cup S_p$); n_i is the unit vector normal to S; u_i^0 and p_i^0 are the boundary data. Again, in the case of $\tau = 0$, p_i reduces to t_i, the boundary surface tractions, i.e., $p_i = \sigma_{ij}n_j = t_i$ when $\tau = 0$. The mixed boundary conditions on S_u and S_p are sufficient but not necessary conditions on S, and are compatible with the field equations.

Initial conditions may also be given at a reference time (e.g., $t = 0$) in the following form:

$$\text{at } t = 0\colon \quad u_i = u_{i0}; \qquad \dot{u}_i = \dot{u}_{i0} . \tag{7; 8}$$

It is noted that the initial values of accelerations $\ddot{u}_i$ and body forces b_i cannot be assigned arbitrarily, but should be made compatible with equations (7)–(8) and the basic dynamic equations.

The displacements u_i are adopted as the primary variables of the physical system at hand. Equations (1)–(8) thus constitute the primary problem for the dynamic response of a standard linear viscoelastic solid.

2.2. Sensitivity analysis

In shape optimization (or identification) problems the shape configuration of a structure is not prescribed a priori, but is determined along with the primary variables of the system, in order to achieve a set of physical objectives (or to satisfy the so-called identification conditions). In such problems, system's response functionals are usually extremized with respect to shape variations subject to the primary problem constraints (cf. equations (1)–(8)), as well as other behavioral and/or side constraints which may be present in the problem. In a numerical solution of a typical problem, SA expressions of functionals play an important role in providing "search" directions in minimization routines.

Consider a general integral functional I given by

$$I = \int_V \int_T f(u_i, \varepsilon_{ij}, b_i)\, dt\, dV + \int_S \int_T g(u_i, p_i)\, dt\, dS , \tag{9}$$

where the integrals f and g are assumed to be continuous and differentiable functions with respect to their arguments, and $T = [0, t_f]$ represents the "control" period of time, t_f being the final (or terminal) time. The functional I, which is also termed as the general performance criterion in an optimization problem, may serve as the objective function to be minimized, or an integral behavioral constraint functional be satisfied exactly.

The problem of any SA is to compute explicitly derivatives of performance criteria with respect to the decision variables. In the present study, the structural shape itself represents the decision variable. The shape SA for viscoelastic solids under transient response will be performed by using the *material derivative* (MD) *concept* of continuum mechanics (cf. [1–3]).

Since the shape of domain V is treated as the (time-invariant) decision variable, it is convenient to think of V as a continuous medium undergoing a "dynamic" deformation, characterized by pseudo-time parameter ν, as V is transformed under a transformation in the shape SA. Thus, a point x_i in V (at $\nu = 0$) moves to the point x_i^ν in the varied domain V^ν, given by

$$x_i^\nu = x_i + \nu V_i ,$$

where the deformation "velocity" field $V_i = V_i(x_j)$ is time independent and is defined in the whole space, representing the rate of deformation (i.e., variation) in the shape SA. The MD of a continuously differentiable function $w(x_i, t)$ can be given by

$$\mathrm{D}w = w' + w_{,k}V_k\,, \tag{10}$$

where Dw and w' represent the MD and the partial derivative of w (whith respect to v), respectively, and are defined as

$$\mathrm{D}w(x_i, t) = \lim_{v\to 0} \frac{w^v(x_i + vV_i, t) - w(x_i, t)}{v}$$

and

$$w'(x_i, t) = \lim_{v\to 0} \frac{w^v(x_i, t) - w(x_i, t)}{v}.$$

It is noted that partial derivatives with respect to x_i and v commute with each other.

General MD formulas may now be given for volume-time and surface-time integrals. Thus, an integral in $V \times T$ may be defined in terms of a differentiable function $w(x_i, t)$ as follows:

$$\psi_1 = \int_V \int_T w \,\mathrm{d}t\,\mathrm{d}V. \tag{11a}$$

The MD of ψ_1 can be shown to be

$$\mathrm{D}\psi_1 = \int_V \int_T w' \,\mathrm{d}t\,\mathrm{d}V + \int_S \int_T wV_n \,\mathrm{d}t\,\mathrm{d}S\,, \tag{11b}$$

where V_n is the normal component of V_i, and it has been assumed that there do not exist discontinuities of w within V. A general integral may also be defined over $S \times T$ for smooth boundary surfaces as follows:

$$\psi_2 = \int_S \int_T w \,\mathrm{d}t\,\mathrm{d}S \tag{12a}$$

whose MD is given in the following form:

$$\mathrm{D}\psi_2 = \int_S \int_T [w' + (w_{,n} + Hw)\,V_n]\,\mathrm{d}t\,\mathrm{d}S + \oint_\Gamma \int_T [\![w]\!]V_\mu \,\mathrm{d}t\,\mathrm{d}\Gamma\,, \tag{12b}$$

where H is the curvature of the boundary S in R^2 and twice the mean surface curvature of S in R^3 and $(.)_{,n}$ denotes the normal derivative of (.) on S. Any discontinuity of w across the boundary surface curve $\Gamma \in S$, indicated by $[\![w]\!]$ is taken care of by the last integral over $\Gamma \times T$ in equation (12b), V_μ is the component of V_i on S, normal to Γ and tangent to S.

The shape SA procedure for dynamically loaded structures using the MD concept is outlined in [4], in which no variational principles are involved. The procedure consists of i) augmenting the performance criterion with the equations of motion by introducing adjoint displacements u_i^*, ii) integrating by parts in order to reduce the order of differentiations, iii) taking the MD of the augmented functional by the given formulae, iv) substituting the partial derivative forms of the constitutive equations, and boundary and initial conditions, v) integrating by parts again, now in order to increase the order of differentiations, and finally vi) defining and satisfying a suitable adjoint problem.

Some of the steps of the above SA procedure may be explained as follows:

Integration by parts of the augmented functional $\tilde{I}$ given by

$$\tilde{I} = I + \int_V \int_T [u_i^*(\sigma_{ij,j} + b_i - \varrho\ddot{u}_i) + \tau u_i^*(\dot{\sigma}_{ij,j} + \dot{b}_i - \varrho\dddot{u}_i)]\,\mathrm{d}t\,\mathrm{d}V$$

leads to the following expression

$$\tilde{I} =$$
$$\int_V \int_T [f - p_{ij}u_{i,j}^* + b_iu_i^* + (\varrho\dot{u}_i + \tau\varrho\ddot{u}_i - \tau b_i)\,\dot{u}_i^*]\,\mathrm{d}t\,\mathrm{d}V + \int_S \int_T (g + p_iu_i^*)\,\mathrm{d}t\,\mathrm{d}S + \int_V [(\tau b_i - \varrho\dot{u}_i - \tau\varrho\ddot{u}_i)\,u_i^*]_T\,\mathrm{d}V.$$

Using the MD formulae (11)–(12) the MD of $\tilde{I}$ can be expressed in the following form:

$$\begin{aligned}
\mathrm{D}\tilde{I} = \int_V \int_T \bigg[& \frac{\partial f}{\partial u_i} u_i' + \frac{\partial f}{\partial \varepsilon_{ij}} u_{i,j}' + \left(\frac{\partial f}{\partial b_i} + u_i^* - \tau \dot{u}_i^*\right) b_i' - p_{ij} u_{i,j}^{*\prime} \\
& - u_{i,j}^* p_{ij}' + b_i u_i^{*\prime} + \varrho \dot{u}_i^* \dot{u}_i' + (\varrho \dot{u}_i - \tau b_i) \dot{u}_i^{*\prime} + \tau \varrho \ddot{u}_i \dot{u}_i^{*\prime} + \tau \varrho \dot{u}_i^* \ddot{u}_i' \bigg] \,\mathrm{d}t \,\mathrm{d}V \\
& + \int_S \int_T \left[\frac{\partial g}{\partial u_i} u_i' + p_i u_i^{*\prime} + \left(\frac{\partial g}{\partial p_i} + u_i^*\right) p_i'\right] \mathrm{d}t \,\mathrm{d}S \\
& + \int_S \int_T [f - p_{ij} u_{i,j}^* + b_i u_i^* + (\varrho \dot{u}_i + \tau \varrho \ddot{u}_i - \tau b_i) \dot{u}_i^* + (g + p_i u_i^*)_{,n} + H(g + p_i u_i^*)] \, V_n \,\mathrm{d}t \,\mathrm{d}S \\
& + \oint_\Gamma \int_T [\![g + p_i u_i^*]\!] \, V_\mu \,\mathrm{d}t \,\mathrm{d}\Gamma + \int_V [(\tau b_i - \varrho \dot{u}_i - \tau \varrho \ddot{u}) \, u_i^{*\prime} + u_i^*(\tau b_i' - \varrho \dot{u}_i' - \tau \varrho \ddot{u}_i')]_T \,\mathrm{d}V \\
& + \int_S [(b_i - \varrho \dot{u}_i - \tau \varrho \ddot{u}_i) \, u_i^*]_T \, V_n \,\mathrm{d}S \,,
\end{aligned}$$

where the definition of p_{ij} has been used. It has been assumed that the body forces depend explicitly on the shape configuration of the solid through some shape parameters, and that by b_i' the variation of b_i with respect to any shape parameters at fixed points of space is understood (see Example problem in Section 4). It is also noted that since the equations of motion relating u_i and b_i to each other have already been introduced using Lagrange multipliers (i.e., adjoint displacements), the partial derivatives u_i' and b_i' are now completely independent quantities.

Partial derivatives of the constitutive equations, and those of the boundary and initial conditions may be substituted into the above equation, where, for example,

$$\text{on } S_u\colon \quad u_i' = \mathrm{D}u_i^0 - u_{i,k} V_k$$

is taken through the use of equations (5) and (10).

After integrating by parts of the $\mathrm{D}\tilde{I}$ expression the coefficients of some of the variations are set equal to zero, thus defining an adjoint problem. The adjoint problem, which corresponds to the original performance criterion and the field equations is thus given by the following terminal time-boundary value problem:

$$\text{in } V\colon \quad \sigma_{ij,j}^* + \partial f/\partial u_i = \sigma \ddot{u}_i^* - \tau \varrho \dddot{u}_i^* \,, \tag{13}$$

$$\text{on } S_u\colon \quad u_i^* = -\partial g/\partial p_i \,, \tag{14}$$

$$\text{on } S_p\colon \quad t_i^* = \sigma_{ij}^* n_j = \partial g/\partial u_i \,, \tag{15}$$

$$\text{at } t = t_f\colon \quad u_i^* = \dot{u}_i^* = \ddot{u}_i^* = 0 \,, \tag{16}$$

where the adjoint "stress" tensor σ_{ij}^* is given by the following:

$$\sigma_{ij}^* = k_0 t_{ij}^* - k_v \dot{t}_{ij}^* - \partial f/\partial \varepsilon_{ij} \tag{17}$$

and

$$t_{ij}^* = \lambda \theta^* \delta_{ij} + \mu(u_{i,j}^* + u_{j,i}^*) \,, \tag{18}$$

where

$$\theta^* = u_{1,1}^* + u_{2,2}^* + u_{3,3}^* \,. \tag{18}$$

The MD (or the total variation) of I with respect to shape configuration can finally be expressed in terms of the primary and adjoint variables as follows:

$$\begin{aligned}
\mathrm{D}I = & \int_V \int_T \left(\frac{\partial f}{\partial b_i} + u_i^* - \tau \dot{u}_i^*\right) b_i' \,\mathrm{d}t \,\mathrm{d}V \\
& + \int_S \int_T [f - p_{ij} u_{i,j}^* + b_i u_i^* - \tau b_i \dot{u}_i^* + \varrho \dot{u}_i \dot{u}_i^* + \tau \varrho \ddot{u}_i \dot{u}_i^* + (g + p_i u_i^*)_{,n} + H(g + p_i u_i^*)] \, V_n \,\mathrm{d}t \,\mathrm{d}S
\end{aligned}$$

$$+ \int_{S_u} \int_T \left(\frac{\partial g}{\partial u_i} - t_i^* \right) (\mathrm{D}u_i^0 - u_{i,k} V_k) \, \mathrm{d}t \, \mathrm{d}S$$

$$+ \int_{S_p} \int_T \left(\frac{\partial g}{\partial p_i} + u_i^* \right) (\mathrm{D}p_i^0 - p_{i,k} V_k) \, \mathrm{d}t \, \mathrm{d}S + \oint_\Gamma \int_T [\![g + p_i u_i^*]\!] \, V_\mu \, \mathrm{d}t \, \mathrm{d}\Gamma$$

$$+ \int_V \{(-\varrho \dot{u}_i^* + \tau \varrho \ddot{u}_i^* - k_v t_{ij,j}^*) (\mathrm{D}u_{i0} - u_{i,k} V_k)$$

$$+ (\varrho u_i^* - \tau \varrho \dot{u}_i^*) (\mathrm{D}\dot{u}_{i0} - \dot{u}_{i,k} V_k) + \tau \varrho u_i^* (\mathrm{D}\ddot{u}_{i0} - \ddot{u}_{i,k} V_k) - \tau u_i^* b_i' \}_{t=0} \, \mathrm{d}V$$

$$+ \int_S [(\varrho \dot{u}_i + \tau \varrho \ddot{u}_i - \tau b_i) u_i^*]_{t=0} \, V_n \, \mathrm{d}S + \int_S k_v [t_{ij}^* (\mathrm{D}u_{i0} - u_{i,k} V_k)]_{t=0} \, n_j \, \mathrm{d}S \,, \tag{19}$$

where the MD forms of the boundary and initial data (e.g., $\mathrm{D}u_i^0$, $\mathrm{D}u_{i0}$, etc.), as well as the partial derivative form of b_i, b_i', are known from their prescribed forms. In particular, when the body forces do not depend on the shape configuration as in conservative loading cases, b_i' should be set equal to zero. Equations (13)–(19) thus constitute the shape SA results for the standard linear viscoelastic solid under transient response. The shape SA of the general performance criterion I can be calculated for any given deformation velocity field V_i by using equation (19).

3. Hereditary integral type of material

In this section, the memory dependent elastic solids of the Boltzmann-Volterra hereditary integral type (i.e., the BV-solid) will be considered.

3.1. Primary problem

The constitutive equations for a homogeneous, isotropic, linear BV-solid may be taken as follows:

$$\sigma_{ij} = \int_0^t \alpha(t - \bar{t}) \, \dot{\bar{\varepsilon}}_{ij} \, \mathrm{d}\bar{t} \,, \tag{20}$$

where t and $\bar{t}$ are time points, and the kernel α incorporates the memory dependence of the material. The superpositioned bar over the field quantities indicates evaluations at $\bar{t}$. It has been assumed for simplicity that the material is free of stresses and strains at the initial time $t = 0$ [7], and that the relaxation function α is continuous and differentiable. It is noted that no specific forms of α will be given in the present study and that the constitutive equations given by equation (20) may reduce to those corresponding to, for example, a Maxwell solid by a proper choice of α.

It is possible to extend the upper limit of the time integration in equation (20) from t to t_f, where $t_f \geqq t$, taking the time integration in the whole $T = [0, t_f]$ interval. The reason for this is that it is advantageous to work with kernel functions which are even in $(t - \bar{t})$. Thus, by introducing a unit Heaviside step function $H(t - \bar{t})$ defined by

$$H(t - \bar{t}) = \begin{cases} 1, & \bar{t} \leqq t \\ 0, & \bar{t} > t \end{cases} \tag{21}$$

and an even kernel function α_e, the constitutive equations (20) can be written as follows:

$$\sigma_{ij} = \int_T \alpha_e(|t - \bar{t}|) \, H(t - \bar{t}) \, \dot{\bar{\varepsilon}}_{ij} \, \mathrm{d}\bar{t} \,, \tag{22}$$

where

$$\alpha(t - \bar{t}) = \alpha_e(|t - \bar{t}|) \, H(t - \bar{t}) \,. \tag{23}$$

The equations of motion, the boundary and initial conditions for the present BV-solid may be taken exactly the same as the ones corresponding to the standard linear (rate-dependent) solid, except for the boundary conditions on S_p, which are replaced by

$$\text{on } S_p\colon \quad t_i = \sigma_{ij} n_j = t_i^0 \,. \tag{24}$$

Thus, (stress) tractions may be imposed on S_p instead of a linear combination of stresses and the rate of stresses as in equation (6).

3.2. Sensitivity analysis

A *general performance criterion* I is now considered given in the form

$$I = \int_V \int_T f(u_i, \varepsilon_{ij}, b_i) \, dt \, dV + \int_S \int_T g(u_i, t_i) \, dt \, dS \,, \tag{25}$$

where p_i has been replaced by t_i in equation (9). For finding the MD of I, the same procedure as in the case of the standard linear solid may be followed, however, with the addition of two steps which seems to be essential in dealing with nonlocal integrals (in time), i.e.,

i) interchanging the order of variables of integration (i.e., t and $\bar{t}$),

ii) renaming the dummy variables of integration (i.e., t and $\bar{t}$).

Using the MD formulae (10)–(12) and following the MD procedure, it is possible to define a suitable *adjoint problem* given by

$$\text{in } V\colon \quad \sigma^*_{ij,j} + \partial f/\partial u_i = \varrho \ddot{u}_i \,, \tag{26}$$

$$\text{on } S_u\colon \quad u_i^* = -\partial g/\partial t_i \,, \tag{27}$$

$$\text{on } S_p\colon \quad t_i^* = \sigma^*_{ij} n_j = \partial g/\partial u_i \tag{28}$$

$$\text{at } t = t_f\colon \quad u_i^* = \dot{u}_i^* = 0 \,, \tag{29}$$

where the *adjoint "constitutive" equations* are expressed as

$$\sigma^*_{ij} = -\frac{\partial f}{\partial \varepsilon_{ij}} - \int_T \alpha_e(|t - \bar{t}|) \, H(\bar{t} - t) \, \dot{\bar{t}}^*_{ij} \, d\bar{t} \,, \tag{30}$$

where t^*_{ij} is given by equation (18).

The MD (or the total variation) of I with respect to shape configuration can finally be given in terms of the primary and adjoint functions and the deformation velocity V_i as follows:

$$\begin{aligned} DI = {} & \int_V \int_T \left(\frac{\partial f}{\partial b_i} + u_i^* \right) b_i' \, dt \, dV \\ & + \int_S \int_T [f - \sigma_{ij} u^*_{i,j} + b_i u_i^* + \varrho \dot{u}_i \dot{u}_i^* + (g + t_i u_i^*)_{,n} + H(g + t_i u_i^*)] \, V_n \, dt \, dS \\ & + \int_{S_u} \int_T \left(\frac{\partial g}{\partial u_i} - t_i^* \right) (Du_i^0 - u_{i,k} V_k) \, dt \, dS \\ & + \int_{S_p} \int_T \left(\frac{\partial g}{\partial t_i} + u_i^* \right) (Dt_i^0 - t_{i,k} V_k) \, dt \, dS + \oint_\Gamma \int_T [\![g + t_i u_i^*]\!] \, V_\mu \, dt \, d\Gamma \\ & + \int_V \varrho \{ u_i^* (D\dot{u}_{i0} - \dot{u}_{i,k} V_k) - \dot{u}_i^* (Du_{i0} - u_{i,k} V_k) \}_{t=0} \, dV + \int_S \varrho [\dot{u}_i u_i^*]_{t=0} \, V_n \, dS \,. \end{aligned} \tag{31}$$

Equations (26)–(31) thus constitute the shape SA results for the BV-solid under transient response.

4. Example

The shape SA results for the rate dependent viscoelastic solids (cf. section 2) will be checked in the case of a Kelvin-Voigt solid by using a one-dimensional example problem. Employing a distributed body force in terms of the fundamental mode of solution it will be possible to compare the SA results with those obtained through a direct differentiation procedure.

Consider a one-dimensional, homogeneous viscoelastic solid of length L and Young's modulus E. Denoting the axial displacement in the rod by u and the axial normal stress by σ, the relevant constitutive equation may be given from equations (1) and (2) by putting $\tau = 0$, $k_0 = 1$ and $k_v = k_v$ as

$$\sigma = E(u_x + k_v \dot{u}_x), \tag{32}$$

where the subscript x denotes differentiation with respect to x.

The primary problem is described by the following equations:

$$0 < x < L: \quad c^2(u_{xx} + k_v \dot{u}_{xx} + b_0 \, e^{-\omega t} \sin \xi_0 x) = \ddot{u}, \tag{33}$$

$$x = 0: \quad u = 0, \tag{34}$$

$$x = L: \quad u_x + k_v \dot{u}_x = 0, \tag{35}$$

$$t = 0: \quad u = \dot{u} = 0, \tag{36}$$

where the system controlling distributed body force has been assumed to be exponentially decaying in time and sinusoidal in space; b_0 is a parameter and

$$c^2 = E/\varrho, \qquad \xi_0 = \pi/2L, \qquad \omega = c\xi_0.$$

The rod is thus taken as clamped and free at $x = 0$ and L, respectively. It is recognized that a simple analytical solution can be obtained to the primary problem in terms of the fundamental characteristic solution due to the nature of the assumed form of the body force distribution, yielding

$$u(x, t) = \frac{c^2 b_0}{2} t^2 \, e^{-\omega t} \sin \xi_0 x, \tag{37}$$

where the viscous damping coefficient k_v has been taken as $k_v = 2/\omega$ (cf. critical damping) for simplicity.

A performance criterion I is defined as

$$I = \frac{1}{2} \int_0^L \int_0^\infty \left(u - \frac{u_d}{b_0 E} b\right)^2 dt \, dx, \tag{38}$$

where u_d represents a given parameter, and b is the body force in the rod, i.e.,

$$b(x, t) = b_0 \, e^{-\omega t} \sin \xi_0 x. \tag{39}$$

It is noted that the control final (or terminal) time t_f has been taken as $t_f \to \infty$, and that I represents the inner product of $(u - u_d b/b_0 E)$ with itself in the case of a square-integrable function space. In an optimization problem, I may play the role of, for example, the objective function, where it is desired to have the system's response (i.e., $u(x, t)$) as proportional to the given body force distribution in the solid.

From equations (9), (13)–(18) and (38) it is found that the adjoint problem corresponding to I is governed by the following equations:

$$0 < x < L: \quad c^2(u^*_{xx} - k_v \dot{u}^*_{xx}) + \frac{1}{\varrho}(u - u_d \, e^{-\omega t} \sin \xi_0 x) = \ddot{u}^*, \tag{40}$$

$$x = 0: \quad u^* = 0, \tag{41}$$

$$x = L: \quad u^*_x - k_v \dot{u}^*_x = 0, \tag{42}$$

$$t \to \infty: \quad u^* = \dot{u}^* = 0, \tag{43}$$

which has a solution for the adjoint displacement u^* given by

$$u^*(x, t) = \frac{1}{16\varrho\omega^2\xi_0^2}(3b_0 - 4u_d\xi_0^2 + 4b_0\omega t + 2b_0\omega^2 t^2) \, e^{-\omega t} \sin \xi_0 x. \tag{44}$$

The length of the viscoelastic rod will be varied by an amount δL at $x = L$, while no variation will take place at $x = 0$, hence

$$\text{at } x = 0: \quad V_n = 0; \qquad \text{at } x = L: \quad V_n = \delta L. \tag{45 a; b}$$

Noting the forms of u and u^* in the rod, it may then be shown that the MD of I is given from equation (19) as follows:

$$\mathrm{D}I = \int_0^L \int_0^\infty \left(\frac{\partial f}{\partial b} + u^*\right) b' \, dt \, dx + \int_0^\infty (f + bu^* + \varrho \dot{u}\dot{u}^*)_{x=L} V_n \, dt, \tag{46}$$

where f is the integrand of I in equation (38), and the partial derivative of b, b', is simply calculated in terms of δL as

$$b' = \frac{\partial b}{\partial L} \, \delta L \, .$$

Introducing equations (37), (39), (44), (45a) into (46), the MD of I is finally given as follows:

$$\mathrm{D}I = \frac{\delta L}{32\omega^5} (9c^4 b_0^2 - 8c^2 b_0 u_d \omega^2 + 8u_d^2 \omega^4) , \tag{47}$$

where DI represents the total variation of I as a function of δL, as the integrand as well as the limit of integration of I change with L.

It is emphasized at this point that the performance criterion I has not been evaluated explicitly in terms of I to find the variation of I with L. Only the shape SA results of section 2 have been employed, which indicates the practical usage of SA expressions in a multi-dimensional problem.

If, on the other hand, equations (37) and (39) are introduced into (38), I can be expressed explicitly in terms of L as

$$I = \frac{L}{64\omega^2 \xi_0^4} (3b_0^2 - 4u_d b_0 \xi_0^2 + 8u_d^2 \xi_0^4) \tag{48}$$

which by differentiation yields

$$\frac{\mathrm{d}I}{\mathrm{d}L} = \frac{1}{32\omega^5} (9c^4 b_0^2 - 8c^2 b_0 u_d \omega^2 + 8u_d^2 \omega^4) \tag{49}$$

which is the same sensitivity expression of I with respect to L derived previously by the shape SA procedure (cf. equation (47)). Hence, for the simple one-dimensional shape variation problem the SA results have been validated by analytical means in the case of Kelvin-Voigt solid under transient response.

References

1 Choi, K. K.; Haug, E. J.: Shape design sensitivity analysis of elastic structures. J. Struct. Mech. **11** (1983), 231–269.
2 Haug, E. J.; Choi, K. K.; Komkov, V.: Design sensitivity analysis of structural systems. Academic Press, New York 1986.
3 Dems, K.; Mroz, Z.: Variational approach by means of adjoint systems to structural optimization and sensitivity analysis. II: Structure shape variation. Internat. J. Solids Structures **20** (1984), 527–552.
4 Meric, R. A.: Shape design sensitivity analysis of dynamic structures. A.I.A.A. Journal **26** (1988), 206–212.
5 Lekszycki, T.; Olhoff, N.: Optimal design of viscoelastic structures under forced steady-state vibration. J. Struct. Mech. **9** (1981), 363–387.
6 Pierson, B. L.: An optimal control approach to the design of vibrating elastic-viscoelastic sandwich beams. Computat. Meths. Appl. Mech. Eng. **57** (1986), 37–49.
7 Christensen, R. M.: Theory of viscoelasticity. Academic Press, New York 1971.

Received November 23, 1988, revised March 10, 1989

Address: Prof. Dr. R. Alsan Meric, Dept. of Space Sciences and Technology, Faculty of Aeronautics and Astronautics, Istanbul Technical University, Maslak, Istanbul 80626, Turkey

ZAMM · Z. angew. Math. Mech. 70 (1990) 11, 517–526 Akademie-Verlag Berlin

BUJURKE, N. M.; NADUVINAMANI, N. B.; JAGADEESWAR, M.

Porous Rayleigh Step Bearing with Second-Order Fluid

The theoretical analysis of a porous Rayleigh Step bearing lubricated by a second-order fluid is considered. An approximate method for the solution of the governing fluid film equations and Darcy's equations for a porous region is proposed. Compact expressions for the pressure distribution, load capacity and the frictional coefficient are obtained. Calculations of the dimensionless load capacity, frictional force and frictional coefficients are presented for specific values of the material parameters. It is found that the load carrying capacity decreases and the frictional force and frictional coefficients increase compared to that of the solid case. The maximum dimensionless load capacity is found to occur at a slightly smaller step ratio as compared to the solid case. It is of quite interest to note that the limiting forms of the expressions obtained approach the exact forms of the results of the Newtonian case quoted in detail in HUGHES [15].

Die Berechnung einer porösen Rayleigh-Lagerpfanne, geschmiert mit einem Fluid zweiter Ordnung, wird betrachtet. Eine Näherungsmethode für die Lösung der geltenden Fluid-Film-Gleichungen und der Darcyschen Gleichungen für ein poröses Gebiet wird angestrebt. Kompakte Ausdrücke für die Druckverteilung, Belastbarkeit und den Reibungskoeffizienten werden gewonnen. Berechnungen der dimensionslosen Belastbarkeit, Reibungskraft und der Reibungskoeffizienten werden vorgestellt für spezifische Werte der Materialparameter. Es zeigt sich, daß — verglichen mit dem Fall des Festkörpers — die Tragfähigkeit abnimmt und die Reibungskraft und Reibungskoeffizienten zunehmen. Die maximale dimensionslose Belastbarkeit kommt bei einem nur wenig kleineren Stufenwert als bei dem des Festkörperfalls vor. Es ist von großem Interesse, daß die erhaltenen geschlossenen Ausdrücke die exakte Form der Resultate des newtonschen Falls in Näherung enthalten, wie sie ausführlich bei HUGHES [15] *dargestellt werden.*

Рассматривается теоретический анализ пористой опорной подушки смазыванной через жидкость второго порядка. Предполагается приближенный метод для решения преобладающих уравнений жидкой пленки и уравнений Дарси в пористой области. Получаются плотные выражения для распределения давления, для нагрузаемости и для коэффициенты трения. Для специфических значений параметров материала даются вычисления безразмерной нагрузаемости, силы трения и коэффициентов трения. Оказывается, что в сравнении с случаем твердого тела нагрузочная способность уменшается а сила трения и коэффициенты трения увеличиваются. Кроме того оказывается, что максимальная нагрузочная способность встречается на незначительно более маленькое отношение шага в случае твердого тела. Доволбно интересно заметить, что предельные формы полученных выражений приближают строгие формы результатов в ньютоновском случае. Это ссылается подробно ЮЗА *в* [15].

1. Introduction

Recently the study of non-Newtonian fluids has attracted much attention, partly because of practical applications in engineering and industry, like fluid film lubrication, analysis of polymers in chemical engineering, etc.; and partly because of academic interest in varieties of fluid flow problems. The most commonly used lubricants are polymer solutions which are non-Newtonian in their character. In the category of non-Newtonian fluids, second-order fluids have distinct features, such as normal stress effects in addition to possessing a large viscosity.

There have been numerous analytical studies of porous bearings which include work by CAMERON et al. [1], ROULEAU [2], CUSANO [3], MURTI [4], PRAKASH and VIJ [5] and WU [6]. They have confined their studies to Newtonian fluids as lubricants. Rayleigh determined the optimum geometry for a viscous incompressible film. The corresponding bearing which is called the Rayleigh Step bearing, has been analysed by MADAY [7] in order to varify that the optimum slider contains only one step. MALIK et al. [19] have given the finite element analysis of Rayleigh Step porous thrust slider bearings using B. J. Slip conditions. Details regarding work on the Rayleigh Step bearings and other aspects of lubrication can be found in recent reviews by LING [8] and SZERI [9]. RAMANAIAH and DUBEY [10], MAITI [11] and SUKLA and ISA [12] have extended the lubrication study of micropolar fluids for Rayleigh Step bearings and have predicted some advantages over Newtonian lubricants.

So far no attempt has been made to study lubrication characteristics of porous Rayleigh Step bearings with second-order fluid as the lubricant. In this paper an attempt is made to study the lubrication characteristics of second-order fluid by adopting a differential type of constitutive equation, proposed by COLEMAN and NOLL [13], which can account for the normal stress effects, in the fluid film region and Darcy's equations for a porous matrix. It is found that the load capacity decreases and the coefficient of friction increases as compared to the solid case. Further, the geometry leading to the optimal load capacity is found to vary slightly compared with the solid case with pure viscous fluid as lubricant [15].

An approximate analytical method is used for the analysis of the problem and it is worth mentioning that the limiting forms of the expressions obtained here agree exactly with closed form solutions of the classical case [15].

2. Basic equations

The constitutive equation for incompressible homogeneous fluid of second-order based on the postulate of a gradually fading memory is given by [13]

$$s_{ij} = -p\delta_{ij} + \varphi_0 A(1)_{ij} + \varphi_1 A(2)_{ij} + \varphi_2 A(1)_{ik} A(1)_{kj}, \tag{1}$$

where s_{ij} is the stress tensor, p is the pressure and φ_0, φ_1 and φ_2 are material constants. Tensors $A(1)_{ij}$ and $A(2)_{ij}$ are given by

$$A(1)_{ij} = v_{i,j} + v_{j,i} \quad \text{and} \quad A(2)_{ij} = a_{i,j} + a_{j,i} + 2v_{m,i}v_{m,j},$$

where v_i and a_i are components of velocity and acceleration and

$$a_i = \frac{dv_i}{dt} + v_j v_{i,j}.$$

In the present study, the model represented by (1) is considered as an exact model (DUNN and FOSDISK [14]). Earlier this model was shown to be a second order approximate to the response functions of a simple fluid in the sense of retardation by COLEMAN and NOLL [13].

From equation (1), the cartesian components of the stress tensor, in the two-dimensional case, are

$$s_{xx} = -p + 2\varphi_0 \frac{\partial u}{\partial x} + \varphi_1 \left[2 \frac{\partial}{\partial x}\left(u \frac{\partial u}{\partial x} + v \frac{\partial u}{\partial y}\right) + 2\left\{\left(\frac{\partial u}{\partial x}\right)^2 + \left(\frac{\partial v}{\partial x}\right)^2\right\}\right] + \varphi_2 \left[4\left(\frac{\partial u}{\partial x}\right)^2 + \left(\frac{\partial u}{\partial y} + \frac{\partial v}{\partial x}\right)^2\right] \tag{2}$$

$$s_{xy} = \varphi_0 \left(\frac{\partial u}{\partial y} + \frac{\partial v}{\partial x}\right) + \varphi_1 \left[\frac{\partial}{\partial y}\left(u \frac{\partial u}{\partial x} + v \frac{\partial u}{\partial y}\right) + \frac{\partial}{\partial x}\left(u \frac{\partial v}{\partial x} + v \frac{\partial v}{\partial y}\right) + 2\left(\frac{\partial u}{\partial x}\frac{\partial u}{\partial y} + \frac{\partial v}{\partial x}\frac{\partial v}{\partial y}\right)\right] + 2\varphi_2 \left[\left(\frac{\partial u}{\partial x} + \frac{\partial v}{\partial y}\right)\left(\frac{\partial u}{\partial y} + \frac{\partial v}{\partial x}\right)\right], \tag{3}$$

and

$$s_{yy} = -p + 2\varphi_0 \frac{\partial v}{\partial y} + 2\varphi_1 \left[\frac{\partial}{\partial y}\left(u \frac{\partial v}{\partial x} + v \frac{\partial v}{\partial y}\right) + \left(\frac{\partial u}{\partial y}\right)^2 + \left(\frac{\partial v}{\partial y}\right)^2\right] + \varphi_2 \left[4\left(\frac{\partial v}{\partial y}\right)^2 + \left(\frac{\partial u}{\partial y} + \frac{\partial v}{\partial x}\right)^2\right]. \tag{4}$$

The flow of a viscous fluid in a porous matrix is governed by

$$\frac{\partial u_1}{\partial x} + \frac{\partial v_1}{\partial y} = 0. \tag{5}$$

The velocity components u_1 and v_1 in the porous region are given by Darcy's law

$$u_1 = -\frac{k_1}{\varphi_0}\frac{\partial p_1}{\partial x}, \qquad v_1 = -\frac{k_1}{\varphi_0}\frac{\partial p_1}{\partial y}, \tag{6), (7}$$

where p_1 is the pressure in the porous region.

3. Mathematical formulation of the problem

We consider a bearing in which one of the two mutually opposed surfaces has pure tangential sliding motion relative to the other surface and the lower surface of the bearing is a porous bed. The geometry of a two dimensional infinite porous Rayleigh Step slider bearing is shown in Figure 1. It is the junction of two seperate bearings of lengths L_1 in region I ($0 \leqq x \leqq L_1$) and L_2 in region II ($L_1 \leqq x \leqq L$). The film thickness in the entry region I and the exit region II are h_1 and h_2, respectively.

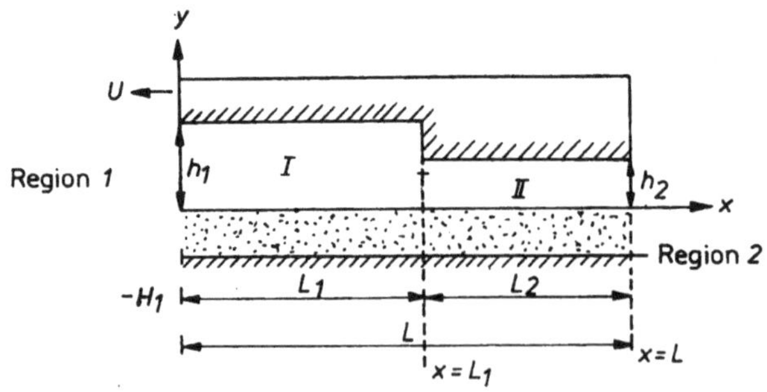

Fig. 1. Porous Rayleigh Step bearing

The momentum and continuity equations of the film region are

$$\frac{\partial}{\partial x}(s_{xx}) + \frac{\partial}{\partial y}(s_{xy}) = 0 \tag{8}$$

$$\frac{\partial}{\partial x}(s_{xy}) + \frac{\partial}{\partial y}(s_{yy}) = 0 \tag{9}$$

and

$$\frac{\partial u}{\partial x} + \frac{\partial v}{\partial y} = 0 \tag{10}$$

respectively, we consider a dimensionless scheme;

$$\varepsilon = \frac{h_2}{L}, \quad \bar{u} = \frac{u}{V}, \quad \bar{v} = \frac{v}{V}, \quad \bar{x} = \frac{x}{L}, \quad \bar{y} = \frac{y}{\varepsilon L}, \quad \bar{p} = \frac{\varepsilon^2 L p}{\varphi_0 V}, \tag{11}$$

where L is the characteristic length and V is the characteristic velocity. Since the minimum thickness of the fluid film is very small compared with the length, L, the dimensionless parameter, ε, is a very small quantity. Substituting (11) into equation (8), and (9), collecting the dominating terms and using lubrication approximation, we obtain

$$\frac{\partial p}{\partial x} = \varphi_0 \frac{\partial^2 u}{\partial y^2} + \varphi_1 \left(\frac{\partial^2 u}{\partial y^2}\frac{\partial u}{\partial x} + 4\frac{\partial u}{\partial y}\frac{\partial^2 u}{\partial x \partial y} + \frac{u \partial^3 u}{\partial x \partial y^2} + \frac{\partial^2 v}{\partial y^2}\frac{\partial u}{\partial y} + v\frac{\partial^3 u}{\partial y^3}\right) + 2\varphi_2 \frac{\partial^2 u}{\partial x \partial y}\frac{\partial u}{\partial y} \tag{12}$$

and

$$\frac{\partial p}{\partial y} = 2(2\varphi_1 + \varphi_2)\frac{\partial u}{\partial y}\frac{\partial^2 u}{\partial y^2}. \tag{13}$$

With reference to Fig. 1, the boundary of the upper surface is given by

$$h = h_i \, (i = 1 \text{ for } 0 \leqq x \leqq L_1) \text{ in region I}, \quad (i = 2 \text{ for } L_1 \leqq x \leqq L) \text{ in region II}, \tag{14}$$

$L = L_1 + L_2$ is the characteristic length of the bearing. The boundary conditions for region (1) are

$$u = u_1 \quad \text{at} \quad y = 0, \tag{15a}$$

$$u = -U \quad \text{at} \quad y = h_i \quad (i = 1, 2), \tag{15b}$$

$$v = 0 \quad \text{at} \quad y = h_i \quad (i = 1, 2), \tag{15c}$$

$$v = v_1 \quad \text{at} \quad y = 0 \tag{15d}$$

and

$$\int_0^{h_i} (p - \tau_{xx})\, dy = 0 \quad \text{at} \quad x = 0 \quad \text{and} \quad x = L. \tag{15e}$$

Also, using the continuity of the total stress at the step ($x = L_1$), we have

$$\int_0^{h_1} (p - \tau_{xx})\, dy = \int_0^{h_2} (p - \tau_{xx})\, dy = 0, \tag{15f}$$

where p is the pressure in the fluid film region. The boundary conditions for region 2 are

$$p_1 = 0 \quad \text{at} \quad x = 0 \quad \text{and} \quad x = L, \tag{15g}$$

$$\left.\frac{\partial p_1}{\partial y}\right|_{y = -H_1} = 0, \tag{15h}$$

$$p(x, 0) = p_1(x, 0). \tag{15i}$$

Equation (15i) implies that there is a continuity of pressure at the interface $y = 0$.

4. Solution of the problem

Using equations (6) and (7) in equation (5), we get

$$\frac{\partial^2 p_1}{\partial y^2} + \frac{\partial^2 p_1}{\partial x^2} = 0. \tag{16}$$

Integrating equation (16) with respect to y over the porous layer thickness and using equation (15h), we obtain

$$\left.\frac{\partial p_1}{\partial y}\right|_{y=0} = -\int_{-H_1}^{0} \frac{\partial^2 p_1}{\partial x^2}\,dy. \tag{17}$$

Assuming the porous layer thickness H_1 to be very small and using equation (15i), equation (17) reduces to

$$\left.\frac{\partial p_1}{\partial y}\right|_{y=0} = -H_1 \frac{\partial^2 p_1}{\partial x^2} = -H_1 \frac{\partial^2 p}{\partial x^2}. \tag{18}$$

This approximation does not produce any significant error as the porous layer thickness is very small [5]. From equations (15i) and (18), we have

$$u_1|_{y=0} = -\frac{k_1}{\varphi_0}\left(\frac{\partial p_1}{\partial x}\right)_{y=0} = -\frac{k_1}{\varphi_0}\frac{\partial p}{\partial x} \tag{19}$$

$$v_1|_{y=0} = -\frac{k_1}{\varphi_0}\left(\frac{\partial p_1}{\partial x}\right)_{y=0} = -\frac{k_1}{\varphi_0}\left(-H_1\frac{\partial^2 p}{\partial x^2}\right) = \frac{k_1}{\varphi_0} H_1 \frac{\partial^2 p}{\partial x^2}. \tag{20}$$

The solution of equations (10), (12) and (13) is attempted in the form

$$u(x, y) = m(x)y^2 + n(x)y + u_0, \tag{21}$$

where $m(x)$ and $n(x)$ are functions of x to be determined. Using equation (10) and the boundary conditions (eqns. (15a)–(15d)), we obtain

$$u(x, y) = m(x)y^2 + n(x)y + u_1 \tag{22}$$

$$v(x, y) = -\frac{dm}{dx}\frac{y^3}{3} - \frac{dn}{dx}\frac{y^2}{2} - \frac{du_1}{dx}y + v_1 \tag{23}$$

where

$$m(x) = \frac{3Uh_i + u_1(3h_i + 6H_1) + C}{h_i^3}, \tag{24}$$

$$n(x) = -\frac{4Uh_i + u_1(4h_i + 6H_1) + C}{h_i^2}. \tag{25}$$

C is an arbitrary constant to be determined. Substituting equations (22)–(25) into (12) and (13), we obtain

$$\frac{\partial p}{\partial x} = 2\varphi_0 m + 4(2\varphi_1 + \varphi_2)\,y^2 \frac{d}{dx}(m^2) + 4(2\varphi_1 + \varphi_2)\,y \frac{d}{dx}(mn) + \left(\frac{3\varphi_1 + 2\varphi_2}{2}\right)\frac{d}{dx}(n^2) + 2\varphi_1 \frac{d}{dx}(mu_1) \tag{26}$$

and

$$\frac{\partial p}{\partial y} = 4(2\varphi_1 + \varphi_2)(2m^2 y + mn). \tag{27}$$

Integration of equations (26) and (27) gives

$$p(x, y) = 2\varphi_0 \int m\,dx + 4(2\varphi_1 + \varphi_2)\,m^2 y^2 + 4(2\varphi_1 + \varphi_2)\,mny + \left(\frac{3\varphi_1 + 2\varphi_2}{2}\right)n^2 + 2u_1 m\varphi_1 + d, \tag{28}$$

where d is a constant to be determined. The average pressure distribution P_i, across the film thickness is given by

$$P_i = \frac{1}{h_i} \int_0^{h_i} p \, \mathrm{d}y$$
$$= 2\varphi_0 \int m \, \mathrm{d}x + 2\varphi_1 u_1 m + \frac{4}{3}(2\varphi_1 + \varphi_2) m^2 h_i^2 + 2(2\varphi_1 + \varphi_2) mnh_i + \left(\frac{3\varphi_1 + 2\varphi_2}{2}\right) n^2 + d \tag{29}$$

and the average total stress is

$$\frac{1}{h_i} \int_0^{h_i} (p - \tau_{xx}) \, \mathrm{d}y = P_i - \hat{\tau}_{xx_i} = 0 \quad \text{at} \quad x = 0 \quad \text{and} \quad x = L\,, \tag{30}$$

where

$$\hat{\tau}_{xx_i} = \frac{1}{h_i} \int_0^{h_i} \tau_{xx} \, \mathrm{d}y = \frac{1}{h_i} \int_0^{h_i} \varphi_2 \left(\frac{\partial u}{\partial y}\right)^2 \mathrm{d}y\,,$$

$$\hat{\tau}_{xx_i} = \varphi_2 \left\{\tfrac{4}{3} m^2 h_i^2 + 2mnh_i + n^2\right\}. \tag{31}$$

Using equation (31) in equation (30) we get

$$\begin{aligned} d = {} & -2\varphi_0 \left(\frac{3Uh_i + C}{h_i^3}\right) x - 2\varphi_1 u_1 \left(\frac{3Uh_i + (3h_i + 6H_1) u_1 + C}{h_i^3}\right) \\ & - \frac{4}{3}(2\varphi_1) \left(\frac{3Uh_i + u_1(3h_i + 6H_1) + C}{h_i^3}\right)^2 h_i^2 \\ & + 2(2\varphi_1) \left(\frac{3Uh_i + (3h_i + 6H_1) u_1 + C}{h_i^3}\right) \left(\frac{4Uh_i + (4h_i + 6H_1) u_1 + C}{h_1^2}\right) h_i \\ & - \frac{3}{2}\varphi_1 \left(\frac{4Uh_i + (4h_i + 6H_1) u_1 + C}{h_i^2}\right)^2 - \frac{4}{3}\varphi_2 \frac{1}{\bar{K}_i}(\bar{K}_i - 1) h_i^2 \left(\frac{3Uh_i + (3h_i + 6H_1) u_1 + C}{h_i^3}\right)^2 \\ & + 2\varphi_2 \frac{1}{\bar{K}_i}(\bar{K}_i - 1) h_i + \left(\frac{3Uh_i + (3h_i + 6H_1) u_1 + C}{h_i^3}\right) \left(\frac{4Uh_i + (4h_i + 6H_1) u_1 + C}{h_i^2}\right) \\ & - \varphi_2 \frac{1}{\bar{K}_i}(\bar{K}_i - 1) \left(\frac{4Uh_i + (4h_i + 6H_1) u_1 + C}{h_i^2}\right)^2, \end{aligned} \tag{32}$$

where

$x = 0$ when $i = 1$ in region I,

$x = L$ when $i = 2$ in region II,

$$\bar{K}_1 = \left[\frac{\bar{H}^3 + 2\bar{k}_1(3\bar{H} + 6\bar{H}_1)}{\bar{H}^3}\right]^{-1}, \qquad \bar{K}_2 = [1 + 2\bar{k}_1(3\bar{H} + 6\bar{H}_1)]^{-1},$$

$$\bar{H} = \frac{h_1}{h_2}, \qquad \bar{k}_1 = \frac{k_1}{h_2^2} \quad \text{and} \quad \bar{H}_1 = \frac{H_1}{h_2}.$$

Substituting equation (32) into equation (29), we obtain

$$\begin{aligned} P_i = {} & \bar{K}_i 2\varphi_0 \left(\frac{3Uh_i + C}{h_i^3}\right) X + \frac{4}{3}\varphi_2 h_i^2 \left(\frac{3Uh_i + (3h_i + 6H_1) u_1 + C}{h_i^3}\right) \\ & - 2\varphi_2 h_i \left(\frac{3Uh_i + (3h_i + 6H_1) u_1 + C}{h_i^3}\right) \left(\frac{4Uh_i + (4h_i + 6H_1) u_1 + C}{h_i^2}\right) \\ & + \varphi_2 \left(\frac{4Uh_i + (4h_i + 6H_1) u_1 + C}{h_i^2}\right)^2, \end{aligned} \tag{33}$$

where

$X = x$ when $i = 1$ in region I,

$X = x - L$ when $i = 2$ in region II.

From the continuity of the total stress at $x = L_1$ given by equation (15f)

$$P_1 - \hat{\tau}_{xx_1} = P_2 - \hat{\tau}_{xx_2} \quad \text{at} \quad x = L_1 . \tag{34}$$

Equation (34) gives

$$C = -3Uh_2\bar{d}_1 , \tag{35}$$

where

$$\bar{d}_1 = \left[\frac{\bar{L}_1\bar{K}_2}{\bar{H}^2} + \bar{L}_2\bar{K}_2\right] \Big/ \left[\frac{\bar{L}_1\bar{K}_1}{\bar{H}^3} + \bar{L}_2\bar{K}_2\right], \qquad \bar{L}_1 = \frac{L_1}{L} \quad \text{and} \quad \bar{L}_2 = \frac{L_2}{L}.$$

Substituting equation (35) into equation (33), and neglecting the terms containing u_1^2 and solving for P, we get

$$P_1 = \bar{K}_1 \frac{6U\varphi_0}{h_2^2}\left(\frac{1}{\bar{H}^2} - \frac{\bar{d}_1}{\bar{H}^3}\right) x + \frac{\varphi_2 U^2}{h_2^2}\left[12\left(\frac{1}{\bar{H}} - \frac{\bar{d}_1}{\bar{H}^2}\right)^2 - 6\left(\frac{1}{\bar{H}} - \frac{\bar{d}_1}{\bar{H}^2}\right)\left(\frac{4}{\bar{H}} - \frac{3\bar{d}_1}{\bar{H}^2}\right) + \left(\frac{4}{\bar{H}} - \frac{3\bar{d}_1}{\bar{H}^2}\right)^2\right]$$

$$- \bar{K}_1\bar{k}_1 \frac{6U^2\varphi_2}{h_2^2}\left(\frac{1}{\bar{H}^2} - \frac{\bar{d}_1}{\bar{H}^3}\right)\left[24\left(\frac{1}{\bar{H}} - \frac{\bar{d}_1}{\bar{H}^2}\right)\left(\frac{\bar{H} + 2\bar{H}_1}{\bar{H}^2}\right) - 6\left(\frac{1}{\bar{H}} - \frac{\bar{d}_1}{\bar{H}^2}\right)\left(\frac{4\bar{H} + 6\bar{H}_1}{\bar{H}^2}\right)\right.$$

$$\left. - 6\left(\frac{4}{\bar{H}} - \frac{3\bar{d}_1}{\bar{H}^2}\right)\left(\frac{\bar{H} + 2\bar{H}_1}{\bar{H}^2}\right) + 2\left(\frac{4}{\bar{H}} - \frac{3\bar{d}_1}{\bar{H}^2}\right)\left(\frac{4\bar{H} + 6\bar{H}_1}{\bar{H}^2}\right)\right](1 - e^{(L_1 - x)/B}) \tag{36}$$

in region I and

$$P_2 = \bar{K}_2 \frac{6U\varphi_0}{h_2^2}(1 - \bar{d}_1)(x - L) + \frac{\varphi_2 U^2}{h_2^2}[12(1 - \bar{d}_1)^2 - 6(1 - \bar{d}_1) - (4 - 3\bar{d}_1) + (4 - 3\bar{d}_1)^2]$$

$$- \bar{K}_2\bar{k}_1 \frac{6U^2\varphi_2}{h_2^2}(1 - \bar{d}_1)[24(1 - \bar{d}_1)(1 + 2\bar{H}_1) - 6(1 - \bar{d}_1)(4 + 6\bar{H}_1) - 6(4 - 3\bar{d}_1)(1 + 2\bar{H}_1)$$

$$+ 2(4 - 3\bar{d}_1)(4 + 6\bar{H}_1)](1 - e^{(L_1 - x)/b_2})$$

in region II where

$$B = -\frac{\bar{k}_1\varphi_2 U}{\varphi_0}\left[24\left(\frac{1}{\bar{H}} - \frac{\bar{d}_1}{\bar{H}^2}\right)\left(\frac{\bar{H} + 2\bar{H}_1}{\bar{H}^2}\right) - 6\left(\frac{1}{\bar{H}} - \frac{\bar{d}_1}{\bar{H}^2}\right)\left(\frac{4\bar{H} + 6\bar{H}_1}{\bar{H}^2}\right)\right.$$

$$\left. - 6\left(\frac{4}{\bar{H}} - \frac{3\bar{d}_1}{\bar{H}^2}\right)\left(\frac{\bar{H} + 2\bar{H}_1}{\bar{H}^2}\right) + 2\left(\frac{4}{\bar{H}} - \frac{3\bar{d}_1}{\bar{H}^2}\right)\left(\frac{4\bar{H} + 6\bar{H}_1}{\bar{H}^2}\right)\right]$$

and

$$b_2 = \frac{\bar{k}_1}{\varphi_0}\varphi_2 U[24(1 - \bar{d}_1)(1 + 2\bar{H}_1) - 6(1 - \bar{d}_1)(4 + 6\bar{H}_1) - 6(4 - 3\bar{d}_1)(1 + 2\bar{H}_1)$$

$$+ 2(4 - 3\bar{d}_1)(4 + 6\bar{H}_1)] .$$

The total load carried by the bearing across the width b is given by

$$W = b\int_0^{L_1} (P_1 - \hat{\tau}_{yy_1})\,dx + b\int_{L_1}^{L} (P_2 - \hat{\tau}_{yy_2})\,dx \tag{38}$$

where

$$\hat{\tau}_{yy_i} = \frac{1}{h_i}\int_0^{h_i} (2\varphi_1 + \varphi_2)\left(\frac{\partial u}{\partial y}\right)^2 dy = (2\varphi_1 + \varphi_2)\left(\frac{4}{3}m^2h_i^2 + 2mnh_i + n^2\right).$$

The dimensionless load capacity is,

$$\bar{W} = \frac{Wh_2^2}{\varphi_0 U L^2 b} = 3\left[\bar{K}_1\left(\frac{1}{\bar{H}^2} - \frac{\bar{d}_1}{\bar{H}^3}\right)\bar{L}_1^2 + \bar{K}_2(\bar{d}_1 - 1)\bar{L}_2^2\right] - \frac{2\varphi_1 U}{\varphi_0 L}\left[\left\{12\left(\frac{1}{\bar{H}} - \frac{\bar{d}_1}{\bar{H}^2}\right)^2\right.\right.$$
$$- 6\left(\frac{1}{\bar{H}} - \frac{\bar{d}_1}{\bar{H}^2}\right)\left(\frac{4}{\bar{H}} - \frac{3\bar{d}_1}{\bar{H}^2}\right) + \left.\left(\frac{4}{\bar{H}} - \frac{3\bar{d}_1}{\bar{H}^2}\right)^2\right\}\bar{L}_1 + \{12(1 - \bar{d}_1) - 6(1 - \bar{d}_1)(4 - 3\bar{d}_1)$$
$$\left. + (4 - 3\bar{d}_1)^2\}\bar{L}_2\right] - 6\frac{\varphi_2 U}{\varphi_0 L}\left[\bar{K}_1\bar{k}_1\left(\frac{1}{\bar{H}^2} - \frac{\bar{d}_1}{\bar{H}^3}\right)\left\{24\frac{1}{\bar{H}} - \frac{\bar{d}_1}{\bar{H}^2}\right)\left(\frac{\bar{H} + 2\bar{H}_1}{\bar{H}^2}\right)\right.$$
$$- 6\left(\frac{1}{\bar{H}} - \frac{\bar{d}_1}{\bar{H}^2}\right)\left(\frac{4\bar{H} + 6\bar{H}_1}{\bar{H}^2}\right) - 6\left(\frac{4}{\bar{H}} - \frac{3\bar{d}_1}{\bar{H}^2}\right)\left(\frac{\bar{H} + 2\bar{H}_1}{\bar{H}^2}\right) + 2\left(\frac{4}{\bar{H}} - \frac{3\bar{d}_1}{\bar{H}^2}\right)\left(\frac{4\bar{H} + 6\bar{H}_1}{\bar{H}^2}\right)\right\}\bar{L}_1$$
$$+ \bar{K}_2\bar{k}_1(1 - \bar{d}_1)\{24(1 - \bar{d}_1)(1 + 2\bar{H}_1) - 6(1 - \bar{d}_1)(4 - 6\bar{H}_1) - 6(4 - 3\bar{d}_1)(1 + 2\bar{H}_1)$$
$$+ 2(4 - 3\bar{d}_1)(4 + 6\bar{H}_1)\}\bar{L}_2 + \bar{K}_1\bar{k}_1\left(\frac{1}{\bar{H}^2} - \frac{\bar{d}_1}{\bar{H}^3}\right)\left\{24\left(\frac{1}{\bar{H}} - \frac{\bar{d}_1}{\bar{H}^2}\right)\left(\frac{\bar{H} + 2\bar{H}_1}{\bar{H}^2}\right)\right.$$
$$+ 6\left(\frac{1}{\bar{H}} - \frac{\bar{d}_1}{\bar{H}^2}\right)\left(\frac{4\bar{H} + 6\bar{H}_1}{\bar{H}^2}\right) - 6\left(\frac{4}{\bar{H}} - \frac{3\bar{d}_1}{\bar{H}^2}\right)\left(\frac{\bar{H} + 2\bar{H}_1}{\bar{H}^2}\right)$$
$$+ 2\left(\frac{4}{\bar{H}} - \frac{3\bar{d}_1}{\bar{H}^2}\right)\left.\left(\frac{4\bar{H} + 6\bar{H}_1}{\bar{H}^2}\right)\right\}(1 - e^{\bar{L}_1/\bar{B}})\bar{B} + \bar{b}_2\{24(1 - \bar{d}_1)(1 + 2\bar{H}_1) - 6(1 - \bar{d}_1)(4 + 6\bar{H}_1)$$
$$\left. - 6(4 - 3\bar{d}_1)(1 + 2\bar{H}_1) + 2(4 - 3\bar{d}_1)(4 + 6\bar{H}_1)\}\bar{K}_2\bar{k}_1(e^{-\bar{L}_2/\bar{b}_2} - 1)\right]. \tag{39}$$

For $\bar{k}_1 = 0$ and φ_1, $\varphi_2 \to 0$, $\bar{W}$ reduces to Newtonian case [15].

The dimensionless frictional force on the lower surface of the bearing ($y = 0$), is

$$\bar{f}_0 = \frac{f_0 h_2}{\varphi_0 U L} = \left[\left\{\left(\frac{4}{\bar{H}} - \frac{3\bar{d}_1}{\bar{H}^2}\right) + 6\bar{K}_1\bar{k}_1\left(\frac{4}{\bar{H}} + \frac{6\bar{H}_1}{\bar{H}^2}\right)\left(\frac{1}{\bar{H}^2} - \frac{\bar{d}_1}{\bar{H}^3}\right)\right\}\bar{L}_1 + \{(4 - 3\bar{d}_1)\right.$$
$$\left. + 6\bar{K}_2\bar{k}_1(4 + 6\bar{H}_1)(1 - \bar{d}_1)\}\bar{L}_2\right]\frac{\varphi_1 U}{\varphi_0 L}\left[\left(\frac{4}{\bar{H}} - \frac{3\bar{d}_1}{\bar{H}^2}\right) + (4 - 3\bar{d}_1)\right]$$
$$+ \frac{6U\varphi_2\bar{k}_1^2}{\varphi_0 L}\left[\bar{K}_1\left(\frac{4}{\bar{H}} + \frac{6\bar{H}_1}{\bar{H}^2}\right)\left(\frac{1}{\bar{H}} - \frac{\bar{d}_1}{\bar{H}^3}\right)\left\{24\left(\frac{1}{\bar{H}} - \frac{\bar{d}_1}{\bar{H}^2}\right)\left(\frac{\bar{H} + 2\bar{H}_1}{\bar{H}^2}\right)\right.\right.$$
$$- 6\left(\frac{1}{\bar{H}} - \frac{\bar{d}_1}{\bar{H}^2}\right)\left(\frac{4\bar{H} + 6\bar{H}_1}{\bar{H}^2}\right) - 6\left(\frac{4}{\bar{H}} - \frac{3\bar{d}_1}{\bar{H}^2}\right)\left(\frac{1}{\bar{H}} + \frac{2\bar{H}_1}{\bar{H}^2}\right) + 2\left(\frac{4}{\bar{H}} - \frac{3\bar{d}_1}{\bar{H}^2}\right)\left.\left(\frac{4\bar{H} + 6\bar{H}_1}{\bar{H}^2}\right)\right\}(1 - e^{\bar{L}_1/\bar{B}})$$
$$+ \bar{K}_2(4 + 6\bar{H}_1)(1 - \bar{d}_1)(1 - e^{-\bar{L}_2/\bar{b}_2})\{24(1 - \bar{d}_1)(1 + 2\bar{H}_1) - 6(1 - \bar{d}_1)(4 + 6\bar{H}_1)$$
$$\left. - 6(4 - 3\bar{d}_1)(1 + 2\bar{H}_1) + 2(4 - 3\bar{d}_1)(4 + 6\bar{H}_1)\}\right]. \tag{40}$$

For $\bar{k}_1 = 0$ and φ_1, $\varphi_2 \to 0$ equation (40) reduces to Newtonian case [15].

The coefficient of friction $\bar{C}$ can be obtained from

$$\bar{C} = \frac{\bar{f}_0}{\bar{W}} \tag{41}$$

5. Discussion and conclusions

Calculations of the load capacity, $\bar{W}$, frictional force, $\bar{f}_0$, and coefficient of friction, $\bar{C}$, for two second-order fluid samples, A and B (material parameters given in table I) for different values of $\bar{H}$, U, $\bar{L}_1$, $\bar{L}_2$ and k_1 are presented in Figs. 2–8. In all these figures the continuous curves refer to fluid A and the dashed ones to fluid B. From Figs. 2 and 3 we observe that $\bar{W}$ gradually increases with an increase in $\bar{H}$ for both fluids A and B. The variation of $\bar{f}_0$ as a function of $\bar{H}$ is shown in Figs. 4 and 5. In these two cases $\bar{f}_0$ decreases with an increase in U.

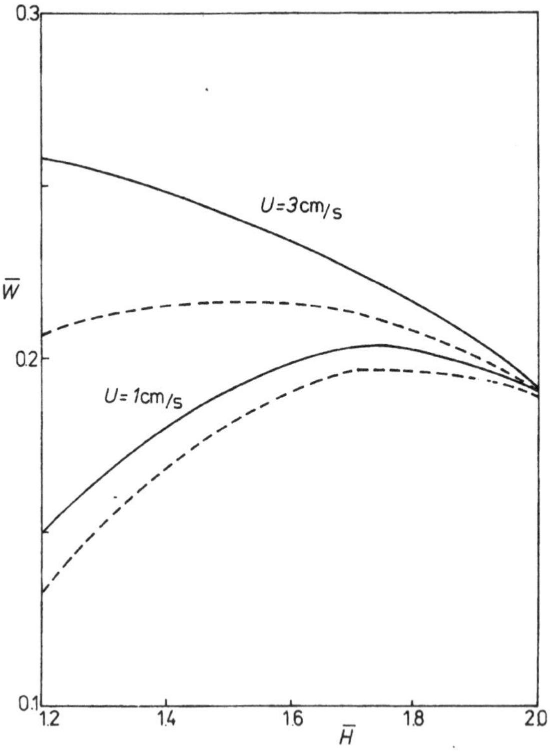

Fig. 2. Variation of $\bar{W}$ with $\bar{H}$ for different values of U for $\bar{L}_1 = 0.7$ and $\bar{L}_2 = 0.3$

Table 1. Material parameters for polyisobutylene in Cetane at 30 °C [18]

Fluid Description	φ_0 (Poise)	φ_1 (Poise-s)	φ_2 (Poise-s)
Fluid A (5.39%)	18.5	−0.3	1.2
Fluid B (5.4%)	18.5	−0.2	1.0

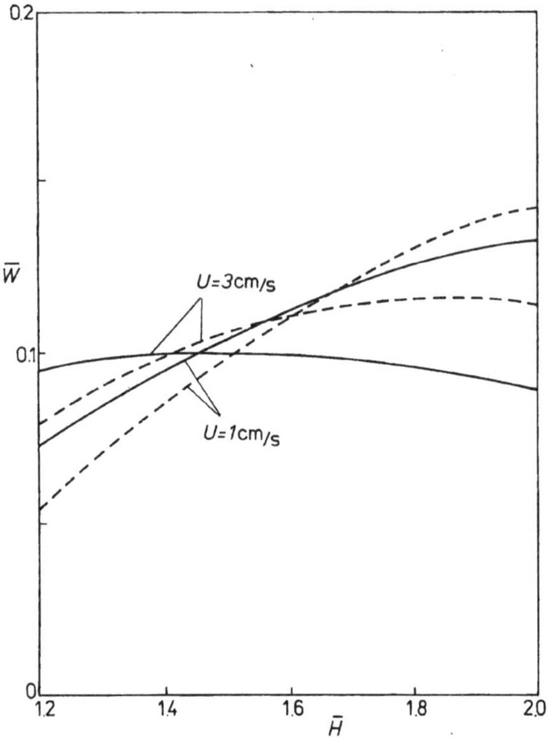

Fig. 3. Variation of $\bar{W}$ with $\bar{H}$ for different values of U for $\bar{L}_1 = 0.9$ and $\bar{L}_2 = 0.1$

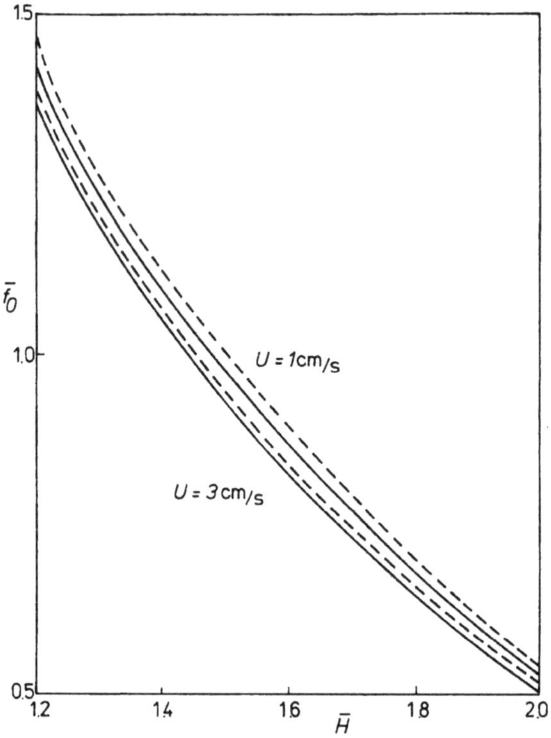

Fig. 4. Variation of $\bar{f}_0$ with $\bar{H}$ for different values of U for $\bar{L}_1 = 0.7$ and $\bar{L}_2 = 0.3$

The maximum value of $\bar{W}$ ($\bar{W}_{max}$) is attained for $\bar{H} = 1.8$ and $\bar{L}_1 = 0.7$ and $\bar{L}_2 = 0.3$ for $U = 1$ cm/s for both the samples. $\bar{W}_{max}$ is larger for fluid A than that for fluid B. As U increases the point of maximum load is shifting towards the entry region. For the solid non Newtonian case, $\bar{W}_{max}$ is found for $\bar{H} = 1.83$ and $\bar{L}_1 = 0.7$ and $\bar{L}_2 = 0.3$. Thus a decrease of 2.4% is found in the present case compared to the solid case. For the solid case, Newtonian fluid of the same viscosity, $\bar{W}_{max}$ is found [15] for $\bar{H} = 1.87$ and $\bar{L}_1 = 0.721$ and $\bar{L}_2 = 0.279$. In the Newtonian porous case a decrease of 0.85% for $\bar{W}_{max}$ is found. Further, it is of interest to note that for optimal $\bar{W}$, the bearing lengths and step height ratio are slightly smaller in our analysis compared with the Newtonian case.

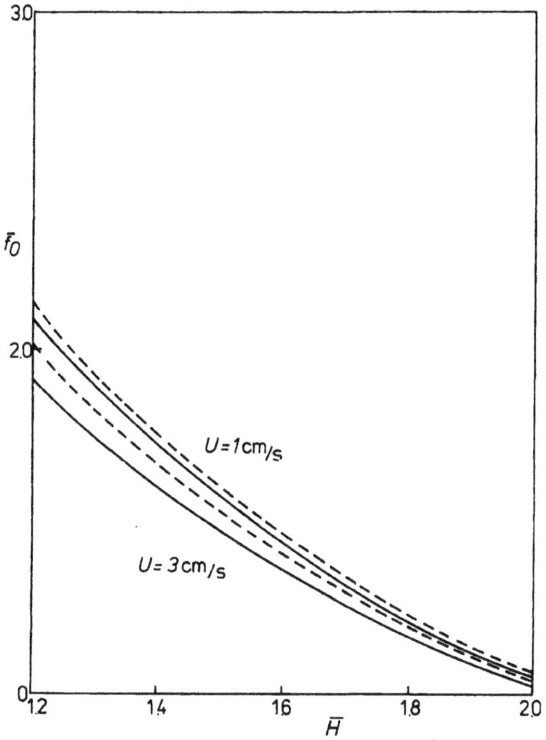

Fig. 5. Variation of $\bar{f}_0$ with $\bar{H}$ for different values of U for $\bar{L}_1 = 0.9$ and $\bar{L}_2 = 0.1$

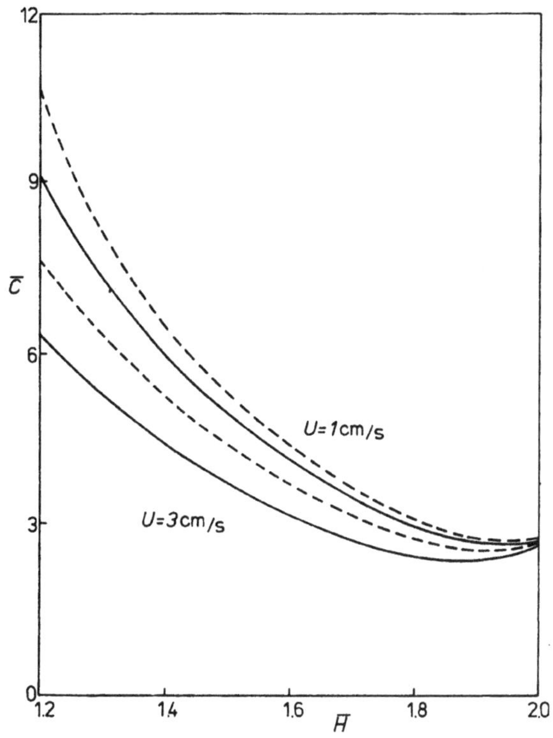

Fig. 6. Variation of $\bar{C}$ with $\bar{H}$ for different values of U for $\bar{L}_1 = 0.7$ and $\bar{L}_2 = 0.3$

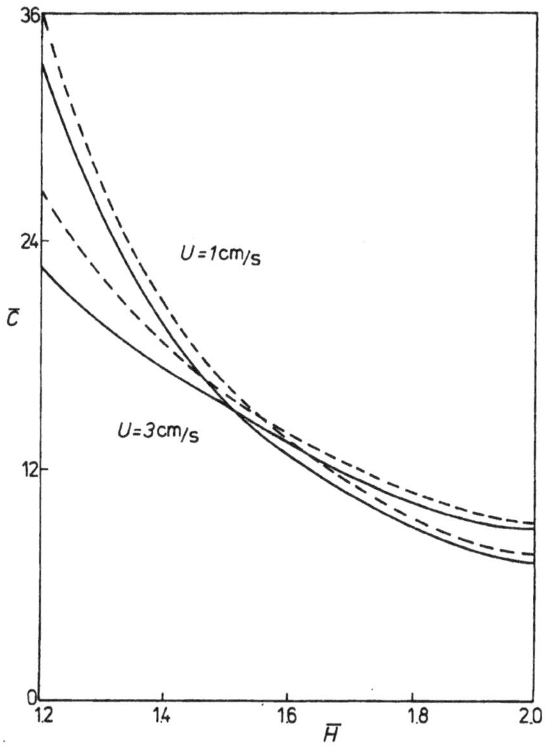

Fig. 7. Variation of $\bar{C}$ with $\bar{H}$ for different values of U for $\bar{L}_1 = 0.9$ and $\bar{L}_2 = 0.1$

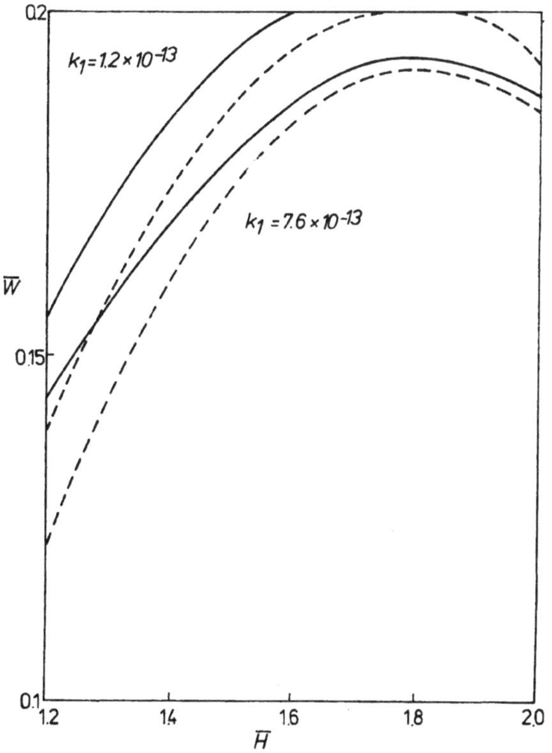

Fig. 8. Variation of $\bar{W}$ with $\bar{H}$ for different values of k_1 for $\bar{L}_1 = 0.7$, $\bar{L}_2 = 0.3$ and $U = 1$ cm/s

Figs. 6 and 7 show the variation of $\bar{C}$ with $\bar{H}$. In all cases $\bar{C}$ decreases with an increase of U a desirable characteristic of lubricants [16], [17]. Fig. 8 shows the variation of bearing load with $\bar{H}$ for different values of k_1. As k_1 decreases the load carrying capacity of the bearing increases significantly.

References

1 Cameron, A.; Morgon, V. T.; Stainsby, A. E.: Proc. Inst. Mech. Eng. London **176** (1962), 761.
2 Roleau, W. T.: J. Basic Eng. **85** (1963) 1, 123.
3 Cusano, C.: J. Lubr. Technol. **94** (1972) 1.
4 Murti, P. R. K.: Wear **19** (1972), 113.
5 Prakash, J.; Vij, S. K.: Wear **24** (1973), 309–322.
6 Wu, H.: Trans. ASME **F94** (1972), 64–68.
7 Maday, C. J.: J. Lubr. Tech. **90** (1968), L240.
8 Ling, F. F.: SIAM. Rev. **28** (1986), L343.
9 Szeri, J.: Tribology, March 1, 1986, Trib-47.
10 Ramanaiah, G.; Dubey, J. N.: Wear **42** (1977), L1.
11 Maiti, G.: Jpn. J. Appl. Physics. **12** (1973), L1058.
12 Shukla, J. B.; Isa, M.: Wear **30** (1974), L51.
13 Coleman, B. D.; Noll, W.: Arch. Ration. Mech. Anal. **6** (1960), L355.
14 Dunn, J. E.; Fosdisk, R. L.: Arch. Ration. Mech. Anal. **56** (1974), 191–252.
15 Hughes, W. F.: An introduction to viscous flow. Carnegie Mellan University 1979, Chap. 3, p. 77.
16 Kragelsky, I. V.; Alisin, V. V.: Friction, wear, lubrication tribology hand book. Mir Publishers, Moscow 1982, Vol. 3, Chap. 24, p. 99.
17 Allen, S. J.; Kline, K. A.: Trans. ASME J. Appl. Mech. (1971), L246.
18 Markovitz, H.: Proc. 4th Int. Congress on Rheology. (Providence RI). Academic Press, New York 1963, Part I, 1965, p. 189.
19 Malik, M.; Sinhasan, R.; Chandra, M.: Wear **76** (1982), 1–13.

Received November 14, 1988

Addresses: Dr. N. M. Bujurke, N. B. Naduvinamani, Reader and Research Scholar, Department of Mathematics, Karnatak University, Dharwad, Pin-580003, India; M. Jagadeeswar, Reader, Department of Mathematics, Karnatak College, Dharwad, Pin-580001, India

SHORT COMMUNICATIONS

ZAMM · Z. angew. Math. Mech. 70 (1990) 11, 527–528 Akademie-Verlag Berlin

IANCU, G.; SCHNACK, E.

Two- and Three-dimensional Shape Optimization with FEM[1])

Introduction

Consider the structural shape optimization problem for minimizing the stress gradient at a traction free surface of an FE-model for a homogeneous, isotropic, linear-elastic continuum:

$$\text{minimize Max}\,[\bar{\sigma}^{\mu}_1(\boldsymbol{b}), \ldots, \bar{\sigma}^{\mu}_{NB}(\boldsymbol{b})]\,, \qquad \mu = 1(1)\,M\,, \tag{1}$$

subjected to:

$$\left.\begin{aligned} v^n_i - d^n_i(\Gamma, \Gamma^+) &\leqq 0 \\ -v^n_i - d^n_i(\Gamma, \Gamma^-) &\leqq 0 \end{aligned}\right\}, \qquad i = 1(1)\,NB\,, \tag{2, 3}$$

$$\bar{\sigma}_i - \tilde{\sigma} \leqq 0\,, \qquad i = 1(1)\,N\,, \tag{4}$$

where NB and N are respectively the number of nodal points on the design boundary Γ and in the domain of the boundary value problem Ω. $\boldsymbol{b}$ is the design vector which contains boundary shape variables. The maximum v. Mises stress of all loading cases $\mu = 1(1)\,M$ is a non-linear objective function. Inequalities (2), (3) and (4) represent non-linear constraints which impose limitations on the boundary variation and stress distribution in Ω. The most commonly used approach for solving the problem (1)–(4) is to apply a standard mathematical programming technique based on discrete or continuum sensitivity analysis. The main difficulty with such methods is that they drastically limit the allowed number of design variables. Most of them give also substantially inaccurate results as has been shown in [1], [2]. The present approach overcomes these difficulties by taking advantage of the character of the problem.

Solution method and numerical tests

In the present formulation the boundary is approximated using the same shape functions assumed for the finite elements boundaries. The design variables are nodal point coordinates on Γ. Using function theory it can be shown that for linear elasticity problems the St. Venant principle applies also to small geometrical disturbances and that the maximum v. Mises stress associated with a boundary notch perturbation behaves monotonically with respect to the normalized curvature of perturbation. A second important effect is the reduction of the stress peak due to the interference between sources of stress concentration (reaction law). These results are used (see also [3]) to solve the shape design problem for the 3D-case. The movement of a boundary approximated by piecewise linear functions in the j-th iteration takes the following form:

$$\delta x^i_j = v^i_j \frac{1}{N_i} \sum_{k=1}^{N_i} \gamma^{k,i}_{j-1} a^{k,i}_{j-1}\,, \tag{5}$$

$$\delta y^i_j = v^i_j \frac{1}{N_i} \sum_{k=1}^{N_i} \gamma^{k,i}_{j-1} b^{k,i}_{j-1}\,, \tag{6}$$

$$\delta z^i_j = v^i_j \frac{1}{N_i} \sum_{k=1}^{N_i} \gamma^{k,i}_{j-1} c^{k,i}_{j-1} \tag{7}$$

with

$$a^{k,i}_{j-1} := (y^{k,i+1}_{j-1} - y^{k,i}_{j-1})(z^{k,i-1}_{j-1} - z^{k,i}_{j-1}) - (y^{k,i-1}_{j-1} - y^{k,i}_{j-1})(z^{k,i+1}_{j-1} - z^{k,i}_{j-1})\,, \tag{8}$$

$$b^{k,i}_{j-1} := (z^{k,i+1}_{j-1} - z^{k,i}_{j-1})(x^{k,i-1}_{j-1} - x^{k,i}_{j-1}) - (z^{k,i-1}_{j-1} - z^{k,i}_{j-1})(x^{k,i+1}_{j-1} - x^{k,i}_{j-1})\,, \tag{9}$$

[1]) Lecture presented at the annual GAMM Conference, Karlsruhe, FRG, March 28–31, 1989, in the Session "Strukturoptimierung I"

$$c^{k,i}_{j-1} := (x^{k,i+1}_{j-1} - x^{k,i}_{j-1})(y^{k,i-1}_{j-1} - y^{k,i}_{j-1}) - (x^{k,i-1}_{j-1} - x^{k,i}_{j-1})(y^{k,i+1}_{j-1} - y^{k,i}_{j-1}) \tag{10}$$

and

$$\gamma^{k,i}_{j-1} := \frac{\sqrt{(a^{k,i}_{j-1})^2 + (b^{k,i}_{j-1})^2 + (c^{k,i}_{j-1})^2}}{(a^{k,i}_{j-1})^2 + (b^{k,i}_{j-1})^2 + (c^{k,i}_{j-1})^2}\,. \tag{11}$$

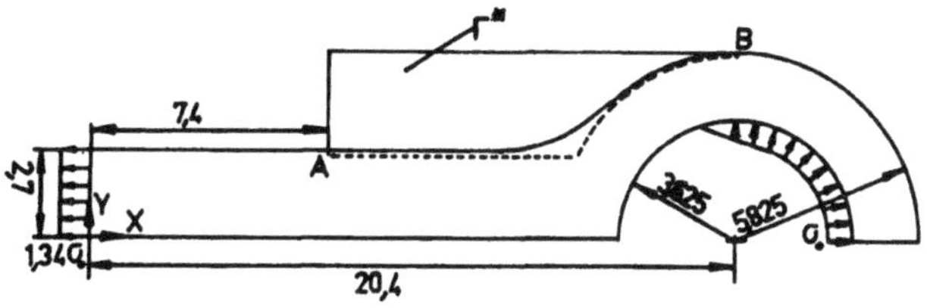

Fig. 1. Connecting rod and variation domain Γ^* with boundary $\Gamma^+ \cup \Gamma^-$. Initial shape and loading

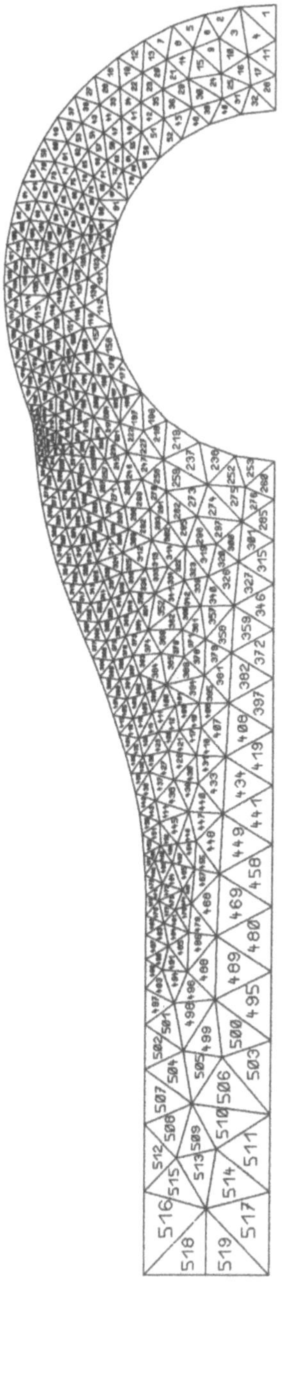

Fig. 2. Optimal shape design of the discretized connecting rod

In equations (5)–(7) N_i is the number of triangle planes abutting the node i whereas v_j^i represents the magnitude of the shifting vector which is computed from its absolute value v_j^{i*} at a node with maximum or minimum v. Mises stress and the sign parameter ζ_{j-1}^i which depends on the stress distribution on Γ according to:

$$v_j^i = \begin{cases} v_j^{i*}\zeta_{j-1}^{i*}, & s = 0, \\ v_j^{i*}\zeta_{j-1}^{i}\left(1 - \dfrac{s}{n_z}\right), & s = 1(1)\,n_z - 1, \\ 0, & s = n_z. \end{cases} \tag{12}$$

Equation (12) describes a smoothing process for a surface divided in n_z zones in order to keep the accuracy of the boundary information at an acceptable level. An example with this procedure for a connecting rod treated as a plane stress problem is given in Figs. 1, 2.

The maximum v. Mises stress which occurs on Γ was reduced here by 63%. Fig. 3 shows a spherical cavity in a body subjected to uniform loading: $\overset{\infty}{\sigma}_x : \overset{\infty}{\sigma}_y : \overset{\infty}{\sigma}_z = 1:2:2$. The maximum v. Mises stress on Γ is $3.36\overset{\infty}{\sigma}_x$. The design surface after 24 iterations approximates an axisymmetrical ellipsoid. It yields a maximum stress value of $2.7\overset{\infty}{\sigma}_x$ and is given in Fig. 4.

References

1 Braibant, V.; Fleury, C.: Sensitivity analysis in shape optimal design. In Mota Soares, C. A. (ed.): CAOD: Structural and mechanical systems. NATO ASI Ser. F27, 1987.

2 Haftka, R.; Barthelemy, B.: On the accuracy of shape sensitivity. In Brebbia, C. A.; Hernandez, S. H. (eds.): Computer aided optimum design of structures: Recent advances. Lecture Notes in Engineering 42, Springer-Verlag, Heidelberg etc. 1989.

3 Schnack, E.; Spörl, U.: A mechanical dynamic programming algorithm for structure optimization. Internat. J. Numer. Math. Eng. 23 (1986), 11.

Received July 27, 1980

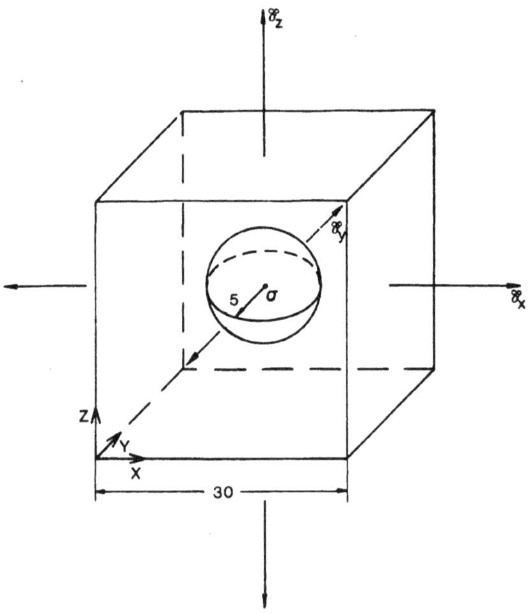

Fig. 3. Cavity in a cube subjected to tension

Address: Dipl.-Phys. Gabriela Iancu; Prof. Dr.-Ing. Eckart Schnack, Institute of Solid Mechanics, Karlsruhe University, Kaiserstraße 12, D-7500 Karlsruhe FRG

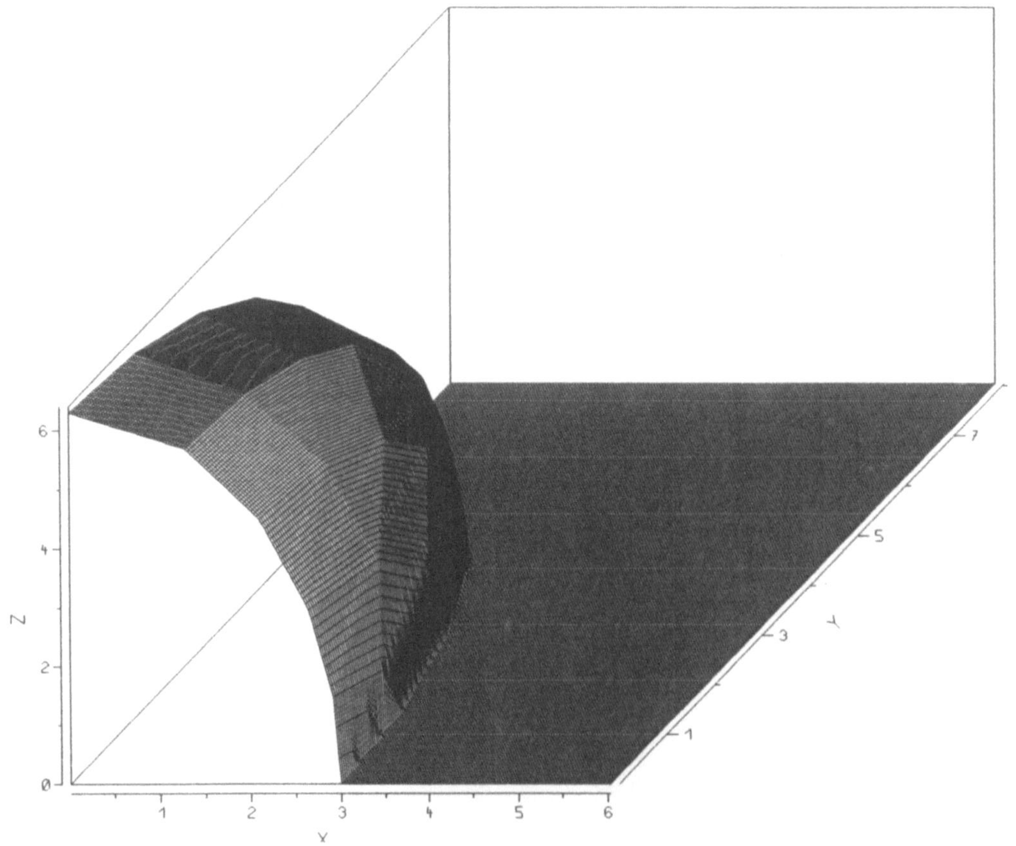

Fig. 4. Approximated ellipsoid

ZAMM · Z. angew. Math. Mech. 70 (1990) 11, 529–530 Akademie-Verlag Berlin

BECKER, I.; SCHNACK, E.

Numerical Calculation of Singularities at Reentrant Edges and Corners for Three Dimensional Crack Problems[1])

Introduction

In LEFM the solution of crack problems is given by the Lamé equation for the displacement $\boldsymbol{u}$. The solution of this elliptical boundary value problem contains singularities at reentrant edges and corners as cracks represents it in three dimensional structures.

In the interior of an elastic body a crack is described by a reentrant edge with an opening angle of 2π. In this case the asymptotic solution is well known. The stress field near the crack tip behaves like $1/\sqrt{r}$. Important parameters to calculate for fracture mechanics are the stress intensity factors given by the coefficients of the $1/\sqrt{r}$ singularity.

At the intersection of a crack front with a free surface the singularity is not known anymore. The intersection point forms a reentrant corner with the crack front as an incoming edge. The domain can be described therefore by a polyhedron in $\mathbb{R}^3$. The following decomposition for the displacement field $\boldsymbol{u}$ in a polyhedral cone with the spherical coordinates (r, θ_j, φ_j) has been proved by M. DAUGE [2] and VON PETERSDORFF [4]:

$$\boldsymbol{u} = \boldsymbol{u}_0 + \chi(r) \sum_{\lambda_k} a_k r^{\lambda_k + 1} \omega_k(\theta_1, \varphi_1) + \sum_{j=1}^{J} \chi_j(\theta_j) \sum_{m,p} e^j_{mp}(r)\, \theta_j^{2p} \boldsymbol{S}^j_m(\theta_j, \varphi_j) + \left[\chi(r) \sum_{l=1}^{L(k)} \sum_{q=0}^{l} r^{\lambda_k+1} \log^q r \hat{\omega}_{klq}(\theta_1, \varphi_1)\right].$$

The *corner singularities* are given by the eigenvalues λ_k while the sum over j describes the *edge singularities*. In special cases there appear additional logarithmic terms. The functions χ are cut-off functions and $\boldsymbol{u}_0$ represents the regular part. For the surface crack this leads to the following expression:

$$\boldsymbol{u} = \boldsymbol{u}_0 + \sum_{\lambda_k} a_k r^{\lambda_k+1} \omega_k(\theta, \varphi) + \varrho^{1/2} \Big(\sum_{\lambda_k} a_k z^{\lambda_k + 1/2} + c(z)\Big) \boldsymbol{f}(\varphi)$$

with $c(z) \to 0$ for $z \to 0$, where the z-axis denotes the crack front and ϱ the distance to it. The displacement field $\boldsymbol{u}$ can be decomposed into a corner singularity given by λ_k and an edge singularity of $\sqrt{\varrho}$-type with a kind of stress intensity factor as coefficient. This stress intensity factor depends on λ_k and tends to zero for $z \to 0$ when the smallest $\lambda_k > -\frac{1}{2}$. Otherwise, it increases to infinity approaching the free surface.

Edge singularity

For the numerical calculation of the stress intensity factors for the edge singularity in case of inner cracks a mixed method is used with trial functions fulfilling the equilibrium and Neumann boundary conditions, see SCHNACK [5] and BECKER et al. [1], realized by the eigenfunctions of the solution for the semi-infinite crack (HARTRANFT, SIH [3]). As a test example the elliptical crack under tension was calculated. Because of symmetry only 1/8 of the domain has to be discretized (Fig. 1). The crack front was approximated by piecewise straight lines by 5 pairs of macro-elements. For the residual domain tetrahedral elements with quadratic displacements were used. The numerical result for the stress intensity factors along the crack front (Fig. 2) agrees well with the analytical values for an infinite body.

Corner singularity

The unknown corner singularity at the intersection point of a crack front with a free surface can be determined by FEM calculations. If logarithmic singularities are absent the asymptotic solution for the stresses has the following form:

$$\sigma_{ij} = r^\lambda f_{ij}(\theta, \varphi).$$

Plotting the stresses versus the distance for constant polar angles θ and φ in a double logarithmic scale one gets a straight line with the unknown eigenvalue as the slope. In Fig. 3 the discretization of 1/8 of a body with a straight crack front coming perpendicular to a free surface is shown. The discretization into 2300 tetrahedrons results in about 11000 equations, which have been solved in about 2 minutes on the vector computer Cyber 205. The stresses on the free surface as a function of the distance from the crack tip along two lines of $\varphi = 45°$ and $\varphi = 60°$ are plotted in a logarithmic scale in Fig. 4. From the slope of the lines one obtains the eigenvalue of the corner singularity:

$$\lambda = -0.40 \pm 0.02.$$

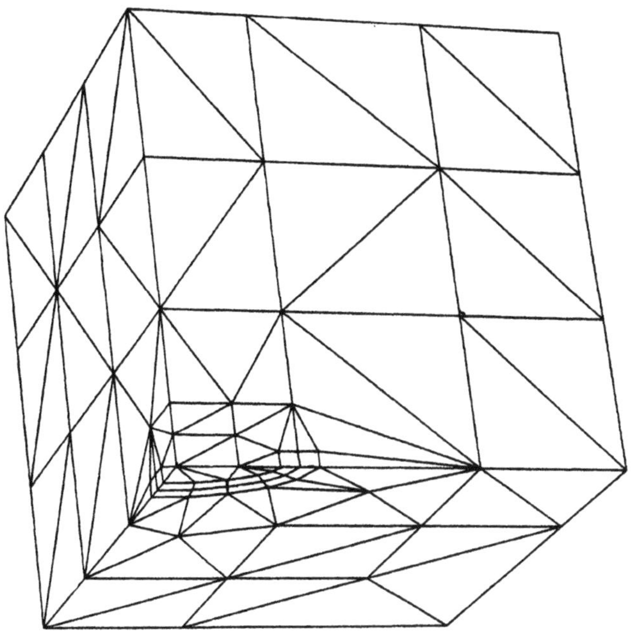

Fig. 1. Discretization for an elliptical crack

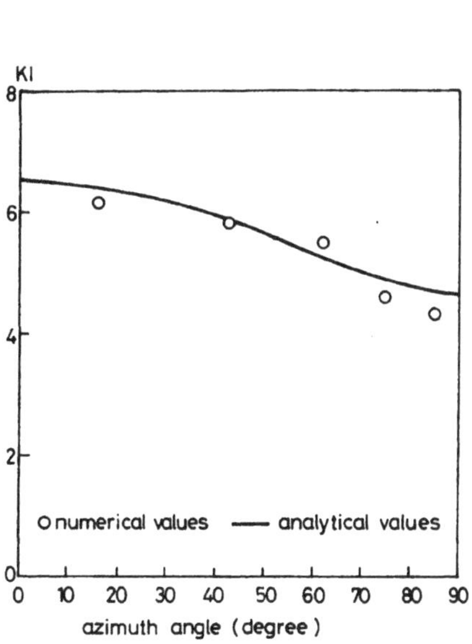

Fig. 2. Stress intensity factor along the crack front

[1]) Lecture presented at the annual GAMM Conference, Karlsruhe, FRG, March 28–31, 1989, in the Session "Einspringende Ecken, singulär gestörte Probleme"

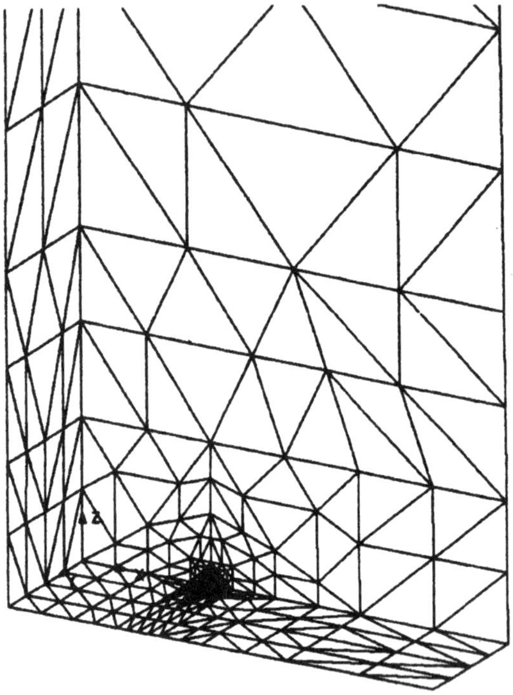

Fig. 3. Discretization for a surface crack

This value is clearly above $-\frac{1}{2}$ which implies a vanishing stress intensity factor at the free surface. A possible logarithmic singularity would result in a curvature in the plot, which cannot be observed in this example.

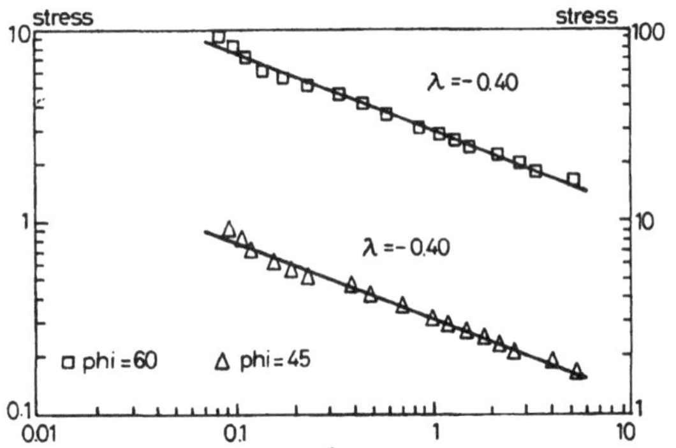

Fig. 4. Plot of the stresses versus the distance from the crack tip

References

1 BECKER, I.; KARAOSMANOĞLU, N.; SCHNACK, E.: Mixed methods with BEM for three dimensional fracture mechanics. In BREBBIA, C. A.; WENDLAND, W. L.; KUHN, G. (eds.): Boundary elements IX 2, 1987, pp. 227–241.

2 DAUGE, M.: Régularités et singularités des solutions de problèmes aux limites elliptiques sur des domaines singulièrs de type à coins. Thèse, Univ. de Nantes 1986.

3 HARTRANFT, R. J.; SIH, G. C.: The use of eigenfunction expansions in the general solution of three dimensional crack problems. J. Math. Mech. 19 (1969), 123–138.

4 v. PETERSDORFF, T.: Elasticity problems in polyhedra – Singularities and numerical approximation with boundary elements. Diss, TH Darmstadt, to appear 1989.

5 SCHNACK, E.; WOLF, M.: Application of displacement and hybrid-stress methods to plane notch and crack problems. Internat. J. Numer. Eng. 12 (1978), 963–975.

Received July 27, 1989

Address: Dipl.-Phys. INGO BECKER, Prof. Dr.-Ing. ECKART SCHNACK, Institute of Solid Mechanics, Karlsruhe University, Kaiserstraße 12, D-7500 Karlsruhe, FRG

ZAMM · Z. angew. Math. Mech. 70 (1990) 11, 530–532 Akademie-Verlag Berlin

ZHANG, SHOUZHU

Fractial Dimension and Fracture Toughness

A fractal model for intergranular and transgranular fractured surfaces of cracked metals is suggested on the basis of a fracture micromechanism. The fractal dimension of the fractured surface is expressed by the grain size and the average distance between two large inclusions. It is shown that the value of the fractal dimension decreases with the increase of the fracture toughness, this is qualitatively consistent with the observations [1, 2, 3].

Introduction

In 1984, MANDELBROT et al. [1] suggested two experimental methods called *Slit Island Analysis* (SIA) and *Fracture Profile Analysis* (FPA) to obtain the fractial dimension of the fractured surfaces of 300 Grand Maraging steel at different temperature of heat treatment. They discovered that the value of the *Fractal Dimension* (FD) decreases smoothly with an increase of the *Impact Energy* (IE) as measured by the standard Charpy impact test. And also PANDE et al. [2] in 1987 measured the fractal dimension of a titanium alloy (Ti-6Al-Zv) with varying zirconium contents by the above mentioned techniques. They found that the FD negatively correlates with *Dynamic Test Energy* (DTE). Recently, MU and LUNG [3] investigated the fractal dimension of the fractured surface of 24 $SiMnCrNi_2Mo$ and 30 $CrMnSiNi_2A$ steels under plane strain conditions and at different test temperatures by means of SIA. They showed that the value of FD decreases smoothly with an increase of the logarithmic value of fracture toughness K_{IC}.

IE, DTE, K_{IC} are the characteristics of material toughness. The negative correlation between the toughness and the value of FD has not been understood satisfactorily yet. MANDELBROT assumed that the correlation must reflect the changes in the microstructure during ageing. From their research, UNDERWOOD et al. [4] tentatively related the changes in FD to the changes in the roughness, shape and confirmation of a curve, but PANDE [2] pointed out that if a high fractal number is supposed to be characteristic for a rough as opposed to a smooth surface we would conclude that the rough surfaces are to be associated with brittle materials. This is contrary to experience. MU and LUNG [3] tried to explain their results in terms of the relation of critical crack extension force to the true total area of the irregular fractured surfaces. However, the explanation would be a failure if we noticed that the yardstick length was less than one.

The better way to understand the correlation is perhaps to investigate it from the micromechanism of fracture. In 1986 LUNG [5] has suggested two kinds of fractal models with which the critical crack extension force was calculated for the intergranular fracture case. It has been determined that the critical crack extension force would rise faster than Hall-Petch's $1/\sqrt{d}$ law with the decrease of grain size due to the increase of the areas of a fractal model of the irregular fractured surfaces. In this paper, we analyse further that the grain size and the average distance between two *Large Inclusions* (LI) or between two *Effective Second Phase Particles* (ESPP), whose sizes are larger than that needed to form voids around them, have an effect on the fracture of metals with fractal model.

The micromechanism of fracture and the fractal model of the fracture surface

The fractal dimension values of FD are at present actually obtained by measuring the length of irregular curves on the fractured surface either by means of SIA or FPA. And we also simply investigate the curves or the fractured surfaces in a two-dimensional system.

The length of the irregular curves is a function of a measuring unit according to the following equation proposed by RICHARDSON [6]:

$$L(\eta) = L_0 \eta^{-(FD-1)}, \tag{1}$$

L_0 is a constant with the dimension of length and FD, the fractal dimension [9], is a constant related to the shape of the linear of form of equation (1):

$$\text{Log}\, L(\eta) = \text{Log}\, L_0 - (\text{FD} - 1)\, \text{Log}\, \eta\,. \tag{2}$$

Irregular curves in general are said (cf. [7]) to possess the property of *self-similitude*, i.e. the curve has the same apparent visual shape and configuration at all magnifications. MANDELBROT [7] provided an alternate method for evaluating FD of "self-similitude" curves,

$$\text{FD} = \log N / \log (1/r)\,, \tag{3}$$

where N is the (new) number of sides on the straight-line segment when a figure is superimposed on the (preceding) segment length. The curves on the fractured surfaces have been assumed to have the "self-similitude" property. In our view, the roughness, shape, and configuration of the fractured surface depend not only on exterior stress but also on the microstructure of the material. The microstructure makes each segment of the cracking to appear in random directions, but the exterior stress keeps the resultant direction of all the segments with the main cracking plan. This inspires us to model the curves on the fractured surfaces in the fractal curves whose generators read as follows (Fig. 1):

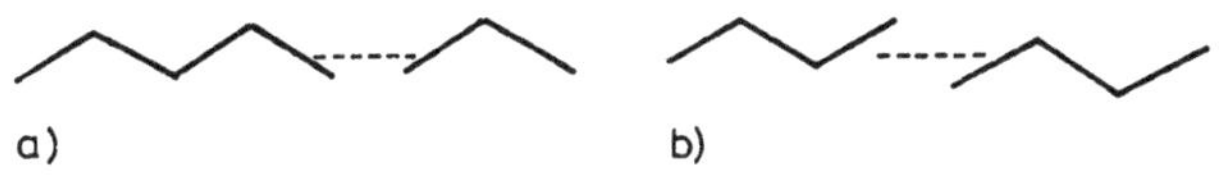

Fig. 1. Generators of fractal curves used to model the curves on a fractured surface. (a) N is equal to an even number. (b) N is equal to an odd number

Each (new) segment is associated with a micro-cracking and has a positive projection on the (preceding) segment on which N micro-crackings are superimposed. The fractal dimension is

$$\left.\begin{aligned} &\text{FD}a = \text{Log}\, N/\text{Log}\,(N \sin \varphi/2)\,, \\ &2 \sin^{-1} (1/\sqrt{N}) < \varphi \leq \pi\,, \qquad N \text{ even};\\ &\text{FD}b = \text{Log}\, N/\text{Log}\,[1 + (N^2 - 1) \sin^2 \varphi/2]^{1/2}\,,\\ &2 \sin^{-1} [1/\sqrt{(N+1)}] < \varphi \leq \pi\,, \qquad N \text{ odd}\,, \end{aligned}\right\} \tag{4}$$

assuming the angles between two adjacent segments are equal. The value of the angle φ depends on the grain configuration. The fractal dimension FD decreases with an increase of the segment number N. This is shown in the table 1.

Table 1. The variation of FD with the changing of the number N of segments and the value of the angle φ

N/φ	2	3	4	5	6	7	8	9
90°	2.0000	1.3652	1.3333	1.2549	1.2398	1.2091	1.2000	1.1833
100°	1.6247	1.2631	1.2380	1.1862	1.1747	1.1538	1.1470	1.1355
120°	1.2619	1.1292	1.1158	1.0932	1.0873	1.0778	1.0743	1.0690
140°	1.0986	1.0526	1.0470	1.0384	1.0360	1.0323	1.0308	1.0278

The relationship between the fractal dimension and the microstructure parameters could be built on the basis of ductile fracture mechanisms and the above analysis. We know from fractography that a lot of voids are firstly created around the ESPP or the LI as a metal deforms plastically, and then increase in size and coalesce into void sheets; these ultimately form a fracture surface. Each plastic pit on the fractured micro-surfaces is corresponding to an ESPP or an LI. The *distance* dt between two pits is determined by the length of the fracture propagation zone. The size and orientation of the grains within dt have an effect on the coalescence of the voids. We therefore hypothesize that the r in the equation (3) is proportional to the ratio of dt to the grain size d, i.e. $r \propto d/\text{dt}$. Two typical forms of intergranular fracture are shown in Fig. 2. And the fractal dimensions of them are

Fig. 2. Two typical forms of intergranular fracture

$$\text{FD}a = \text{Log}\, 2m/\text{Log}\,(3m)\,, \qquad m \text{ even}\,, \tag{5}$$

$$\left.\begin{aligned} &\text{FD}b = \text{Log}\, 2m/\text{Log}\,(3m/2) \quad \text{or}\\ &\text{FD}b = \text{Log}\,(2m+1)/\text{Log}\,(3m/2 + 1/2)\,, \qquad m \text{ odd}\,, \end{aligned}\right\} \tag{6}$$

where $m = 1, 2, 3, \ldots$ is the number of grains on the straight-line segment dt. FD is approximately a linear function of the logarithm values of the grain number m, as it is presented in Fig. 3. Two fractal fractured surfaces in a two dimension system are shown in point line in Fig. 4 on the basis of the model. The first one is the *intergranular fractured surface* and the second one is the *transgranular fractured surface*.

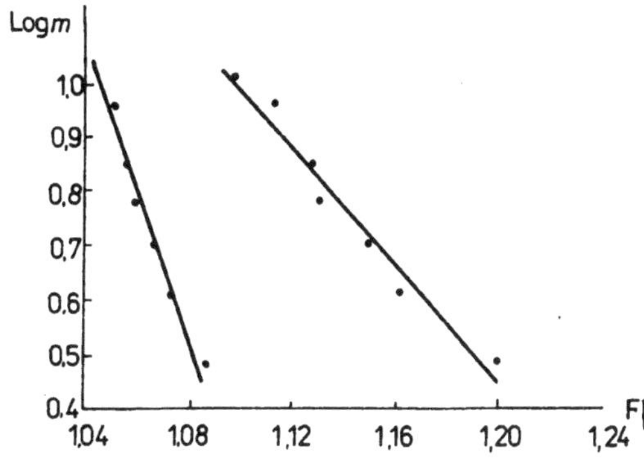

Fig. 3. The fractal dimension number FD versus the logarithmic values of the grain number m

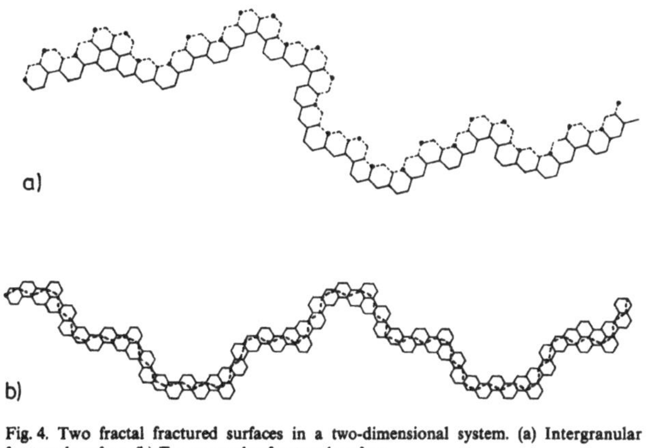

Fig. 4. Two fractal fractured surfaces in a two-dimensional system. (a) Intergranular fractured surface. (b) Transgranular fractured surface

Fractal dimension and fracture toughness

The fractal dimension of the fractured surface can be expressed by the microstructure parameters d and dt according to the above analysis as

$$\text{FD} = \text{Log}\,(c_1\, \text{dt}/d)/\text{Log}\,(c_2\, \text{dt}/d)\,, \tag{7}$$

where c_1, c_2 are the constants associated with the microstructure, such as $c_1 = 1$, $c_2 = \sqrt{3}$ for Fig. 2 (a); $c_1 = 4/\sqrt{3}$, $c_2 = \sqrt{3}$ for

Fig. 2 (b). Some insight into the physical nature of the fractal dimension is afforded by equation (7). The ratio dt/d dominates the values of the fractal dimension. As is well known, the distance dt varies much greater than the grain size d when the test temperature changes. At lower temperature dt becomes shorter because the yield strength of the material increases and the size of ESPP decreases. Consequently, the ratio becomes smaller and FD becomes greater. This is perhaps one of the reasons why the fractal dimension of the fractured surfaces increases with the decrease of the test temperature, which was observed in [3]. We could predict that FD will become greater no matter what method we use (lowering temperature, raising strain ratio or heat treatment) to improve the yield strength. Usually, for a kind of material, the higher the yield strength is, the lower the toughness appears. We therefore come to the conclusion that the fractal dimension of the fractured surface decreases with the increase of the toughness, which is qualitatively consistent with the observations [1, 2, 3].

We can approximately derive the relationship between the fractal dimension FD and (a) the plane strain fracture toughness K_{IC}, or (b) the *Volume Fraction* VF of big inclusions from (7).

a) From [7] we get

$$K_{IC} = (CEX_0 \sigma f \varepsilon f)^{1/2}\,, \tag{8}$$

where C is a constant associated with the hardness and Poission's ratio, E is the Young's modulus, X_0 is the average distance between two large inclusions, i.e. $X_0 = \mathrm{dt}$, σf is the real fracture stress, εf is the real strain under uniaxial pulling condition. Assume that the following relation is still applicable:

$$\sigma f = E \varepsilon f\,; \tag{9}$$

then we obtain

$$\mathrm{FD} = \mathrm{Log}\,[(C_1/C)^{1/2}\, kK_{IC}]/\mathrm{Log}\,[C_2/C)^{1/2} kK_{IC}]\,, \tag{10}$$

from equations (9), (8), (7) and the Hall-Petch's $1/\sqrt{d}$ law, where k is a constant. The relationship between FD and the fracture toughness K_{IC} is the same as that between FD and the grain number m shown in equations (5), (6). For this reason, we can say that FD is approximately a linear function of the logarithmic values of fracture thoughness K_{IC} with reference to Fig. 3. It is just the result obtained in experiment [3].

b) Suppose that there are n inclusions with an average diameter d_1 in a unit volume and all inclusions are large enough to create void around them under an exterior stress. The average distance dt between two inclusions can be written as

$$\mathrm{dt} = (\pi/6)^{1/3}\, d_1/Fv^{1/3}\,, \tag{11}$$

where $Fv = \pi d_1^3\, n/6$ is the volume fraction of the big inclusions. From (11) and (7) we obtain

$$\mathrm{FD} = \mathrm{Log}\,\{C_1[\pi/(6Fv)]^{1/3}\, d_1/d\}/\mathrm{Log}\,\{C_2[\pi/(6Fv)]^{1/3}\, d_1/d\}\,. \tag{12}$$

The fractal dimension FD can be expected to increase with the increase of the volume fraction if the grain size d and the inclusion size d_1 are unchanged.

Fracture phenomena are some of the most intriguing processes in material science. A wide variety of mechanisms plays relevant roles: grain boundaries, microcracks, temperature, anisotropies. Hence, a rich phenomenology exists, and very different physical processes, from cleavage to ductile fracture and fatigue, are included in this field [11]. While the model in this paper is rather simple, its conclusions are qualitatively consistent with the observations [1, 2, 3]. We thus think it may give us one of the clues to understand the observations. Three points should be added:

(a) The fractal fractured surface models shown in Fig. 4 do not mean that the grain size d is the characteristic smallest length scale [9]. The fractographic observations on intergranular fracture indicated that there are some smaller pits on the grain boundaries which result from the smaller precipitated phase particles there, and as for the transgranular fracture, the cracking can not run through the grains in the definite direction since there are many microstructure defects (such as dislocations, cavities, smaller inclusions) in grains. We conjecture that the ratio $r_\alpha d/\mathrm{dt}$ is applicable within the grain size range.

(b) The positive relation between FD and the fracture toughness has been observed (cf. [10]) by the method ISA with ruler much smaller than the previous one [3]. Whether it exposed the deeper level of the microstructure remains for further investigation.

(c) The "self-similitude" property of fractured surface is actually not as simple as that described in the article. To get a good conclusion, the statistical concept might be introduced into considerations [12]. The author will do that next.

Acknowledgements

The author is grateful to Prof. C. W. LUNG for encouraging to finish the work, to Prof. K. L. WANG for hospitality at the University of Science and Technology of China, to Mr. Z. Q. MU and Mr. D. L. CHEN for their helpful discussions.

References

1 MANDELBROT, B. B.; PASSOJA, D. E.; PAULLAY, A. J.: Nature **308** (1984) 19, 721.
2 PANDE, C. S.; RICHARDS, L. E.; LOUAT, N.; DEMPSEY, B. D.; SCHWOEBLE, A. J.; Acta metall. **35** (1987), 1633.
3 MU, Z. Q.; LUNG. C. W.: J. Phys. D: Appl. Phys. **21** (1988), 848.
4 UNDERWOOD, E. E.; BANERJI, K.: Mater. Sci. Engng. **80** (1986), 1.
5 LUNG, C. W. in PIETRONERO, L.; TOSSATTI, E. (eds.): Fractals in physics. 1986, p. 461.
6 RICHARDSON, L. F. quoted in MANDELBROT, B. B.: Fractals: Chance and dimension. W. H. Freeman, San Francisco 1977, p 32.
7 MANDELBROT, B. B.: The fractal geometry of nature. W. H. Freeman San Francisco 1982.
8 OSBORNE; EMBURY: Met. Trans., Granveien **46**, 1360 Nesbru, Norway 1986, p. 26.
10 LUNG, C. W.; MU, Z. Q.: Phys. Rev. (in press).
11 LOUIS, E.; GUINEA, F.: Europhys. lett. **3** (1987) 8, 871.
12 PIETRONERO, L.; ERZAN, A.: EVERTSZ, C.: Physics A **151** (1988), 207.

Received January 12, 1989

Address: Dr. ZHANG SHOUZHU, Fundamental Physics Centre, University of Science and Technology of China, Hefei Anhui 230026, P. R. China
Present address: Dr. SHOUZHU ZHANG, Physics Department of Liaoning University, Shenyang, Liaoning 110036, P.R. China

Bei Fragen zur Produktsicherheit wenden Sie sich bitte an:
If you have any questions regarding product safety,
please contact:

Walter de Gruyter GmbH
Genthiner Straße 13
10785 Berlin
productsafety@degruyterbrill.com